AF678831

THE ROLE OF PEPTIDES AND AMINO ACIDS AS NEUROTRANSMITTERS

PROGRESS IN CLINICAL AND BIOLOGICAL RESEARCH

Series Editors
Nathan Back
George J. Brewer
Vincent P. Eijsvoogel
Robert Grover
Kurt Hirschhorn
Seymour S. Kety
Sidney Udenfriend
Jonathan W. Uhr

RECENT TITLES

Vol 50: **Rights and Responsibilities in Modern Medicine: The Second Volume in a Series on Ethics, Humanism, and Medicine,** Marc D. Basson, *Editor*

Vol 51: **The Function of Red Blood Cells: Erythrocyte Pathobiology,** Donald F. H. Wallach, *Editor*

Vol 52: **Conduction Velocity Distributions: A Population Approach to Electrophysiology of Nerve,** Leslie J. Dorfman, Kenneth L. Cummins, and Larry J. Leifer, *Editors*

Vol 53: **Cancer Among Black Populations,** Curtis Mettlin and Gerald P. Murphy, *Editors*

Vol 54: **Connective Tissue Research: Chemistry, Biology, and Physiology,** Zdenek Deyl and Milan Adam, *Editors*

Vol 55: **The Red Cell: Fifth Ann Arbor Conference,** George J. Brewer, *Editor*

Vol 56: **Erythrocyte Membranes 2: Recent Clinical and Experimental Advances,** Walter C. Kruckeberg, John W. Eaton, and George J. Brewer, *Editors*

Vol 57: **Progress in Cancer Control,** Curtis Mettlin and Gerald P. Murphy, *Editors*

Vol 58: **The Lymphocyte,** Kenneth W. Sell and William V. Miller, *Editors*

Vol 59: **Eleventh International Congress of Anatomy,** Enrique Acosta Vidrio, *Editor-in-Chief.* Published in 3 volumes:
- Part A: **Glial and Neuronal Cell Biology,** Sergey Fedoroff, *Editor*
- Part B: **Advances in the Morphology of Cells and Tissues,** Miguel A. Galina, *Editor*
- Part C: **Biological Rhythms in Structure and Function,** Heinz von Mayersbach, Lawrence E. Scheving, and John E. Pauly, *Editors*

Vol 60: **Advances in Hemoglobin Analysis,** Samir M. Hanash and George J. Brewer, *Editors*

Vol 61: **Nutrition and Child Health: Perspectives for the 1980s,** Reginald C. Tsang and Buford Lee Nichols, Jr., *Editors*

Vol 62: **Pathophysiological Effects of Endotoxins at the Cellular Level,** Jeannine A. Majde and Robert J. Person, *Editors*

Vol 63: **Membrane Transport and Neuroreceptors,** Dale Oxender, Arthur Blume, Ivan Diamond, and C. Fred Fox, *Editors*

Vol 64: **Bacteriophage Assembly,** Michael S. DuBow, *Editor*

Vol 65: **Apheresis: Development, Applications, and Collection Procedures,** C. Harold Mielke, Jr., *Editor*

Vol 66: **Control of Cellular Division and Development,** Dennis Cunningham, Eugene Goldwasser, James Watson, and C. Fred Fox, *Editors*. Published in 2 Volumes.

Vol 67: **Nutrition in the 1980s: Constraints on Our Knowledge,** Nancy Selvey and Philip L. White, *Editors*

Vol 68: **The Role of Peptides and Amino Acids as Neurotransmitters,** J. Barry Lombardini and Alexander D. Kenny, *Editors*

See pages 231–232 for previous titles in this series.

THE ROLE OF PEPTIDES AND AMINO ACIDS AS NEUROTRANSMITTERS

Proceedings of the Third Tarbox Parkinson's Disease Symposium, held on October 16 – 18, 1980, at Texas Tech University Health Sciences Center, Lubbock, Texas

Editors

J. Barry Lombardini
Alexander D. Kenny
Department of Pharmacology and Therapeutics
Texas Tech University Health Sciences Center
Lubbock, Texas

ALAN R. LISS, INC. • NEW YORK

Address all Inquiries to the Publisher
Alan R. Liss, Inc., 150 Fifth Avenue, New York, NY 10011

Copyright © 1981 Alan R. Liss, Inc.

Printed in the United States of America.

Under the conditions stated below the owner of copyright for this book hereby grants permission to users to make photocopy reproductions of any part or all of its contents for personal or internal organizational use, or for personal or internal use of specific clients. This consent is given on the condition that the copier pay the stated per copy fee through the Copyright Clearance Center, Incorporated, 21 Congress Street, Salem, MA 01970, as listed in the most current issue of "Permissions to Photocopy" (Publisher's Fee List, distributed by CCC, Inc.), for copying beyond that permitted by sections 107 or 108 of the US Copyright Law. This consent does not extend to other kinds of copying, such as copying for general distribution, for advertising or promotional purposes, for creating new collective works, or for resale.

THIS BOOK IS NO LONGER THE PROPERTY OF THE U OF C LIBRARY

UNIVERSITY OF CHICAGO LIBRARY

Library of Congress Cataloging in Publication Data

Tarbox Parkinson's Disease Symposium (3rd : 1980 : Texas Tech University Health Sciences Center)

The role of peptides and amino acids as neurotransmitters.

(Progress in clinical and biological research; v. 68)

Includes bibliographies and index.

1. Neurotransmitters–Congresses. 2. Peptides –Physiological effect–Congresses. 3. Amino acids –Physiological effect–Congresses. I. Kenny, Alexander D. II. Lombardini, J. Barry. III. Title. IV. Series. [DNLM: 1. Amino acids–Metabolism–Congresses. 2. Peptides–Metabolism–Congresses. 3. Neuroregulators–Physiology–Congresses. 4. Parkinson's disease–Congresses. WI PR668E v. 68 / WL 359 T179 1980r]

QP364.7.T37 1980 599.01'88 81-8335
ISBN 0-8451-0068-8 AACR2

Contents

Participants

Symposium Chairman

Alexander D. Kenny, PhD
Professor and Chairperson, Department of Pharmacology and Therapeutics, and Director, Tarbox Parkinson's Disease Institute, Texas Tech University Health Sciences Center, Lubbock, Texas

Program Chairman

J. Barry Lombardini, PhD
Associate Professor, Department of Pharmacology and Therapeutics, Texas Tech University Health Sciences Center, Lubbock, Texas

Session Chairmen

Penelope W. Coates, PhD
Associate Professor, Department of Anatomy, Texas Tech University Health Sciences Center, Lubbock, Texas

J. Barry Lombardini, PhD
Associate Professor, Department of Pharmacology and Therapeutics, Texas Tech University Health Sciences Center, Lubbock, Texas

William H. Lyness, PhD
Assistant Professor, Department of Pharmacology and Therapeutics, Texas Tech University Health Sciences Center, Lubbock, Texas

Paul G. Meyer, MD
Chairman and Associate Professor, Department of Medical and Surgical Neurology, and Medical Director, Tarbox Clinic for Parkinson's Disease, Texas Tech University Health Sciences Center, Lubbock, Texas

James H. Pirch, PhD
Professor, Department of Pharmacology and Therapeutics, Texas Tech University Health Sciences Center, Lubbock, Texas

David E. Potter, PhD
Professor, Department of Pharmacology and Therapeutics, Texas Tech University Health Sciences Center, Lubbock, Texas

Invited Speakers

George R. Breese, PhD
Professor, Departments of Psychiatry and Pharmacology, University of North Carolina School of Medicine, Chapel Hill, North Carolina

Jacob M. Hiller, PhD
Research Associate Professor, Department of Psychiatry, New York University Medical Center, New York, New York

Ryan J. Huxtable, PhD
Professor, Department of Pharmacology, University of Arizona, Tucson, Arizona

Graham A.R. Johnston, PhD
Professor and Chairman, Department of Pharmacology, University of Sydney, Sydney, Australia

Kresimir Krnjević, MB, ChB, PhD
Professor and Chairman, Department of Physiology, and Director, Anesthesia Research, McGill University, Montreal, Quebec, Canada

Hugh McLennan, PhD
Professor, Department of Physiology, University of British Columbia, Vancouver, British Columbia, Canada

Earl A. Zimmerman, MD
Professor, Department of Neurology, College of Physicians and Surgeons, Columbia University, New York, New York

Other Participants

Abraham Flemenbaum, MD
Department of Psychiatry, University of Miami, and the Veterans Administration Medical Center, Miami, Florida

Donald N. Franz, PhD
Department of Pharmacology, University of Utah College of Medicine, Salt Lake City, Utah

I.M. Lang, PhD
Department of Physiology, Texas Tech University Health Sciences Center, Lubbock, Texas

T. Celeste Napier, BS
Department of Pharmacology and Therapeutics, Texas Tech University Health Sciences Center, Lubbock, Texas

Vernon J. Perez, PhD
Department of Biochemistry, Texas Tech University Health Sciences Center, and Department of Psychology, Texas Tech University, Lubbock, Texas

Howard K. Strahlendorf, PhD
Department of Medical and Surgical Neurology, Texas Tech University Health Sciences Center, Lubbock, Texas

Jean C. Strahlendorf, PhD
Department of Physiology, Texas Tech University Health Sciences Center, Lubbock, Texas

Contributors

Charles D. Barnes, PhD [161, 197, 217]
Department of Physiology, Texas Tech University Health Sciences Center, Lubbock, TX 79430

George R. Breese, PhD [99]
Departments of Psychiatry and Pharmacology, University of North Carolina School of Medicine, Chapel Hill, NC 27514

Robert L. Casady, PhD [151]
Department of Anatomy, Texas Tech University Health Sciences Center, Lubbock, TX 79430

John C. Eatwell, PhD [151]
Department of Psychology, Texas Tech University, Lubbock, TX 79409

Abraham Flemenbaum, MD [143, 209]
Department of Psychiatry, University of Miami, and the Veterans Administration Medical Center, Miami, FL 33125

Donald N. Franz, PhD [181]
Department of Pharmacology, University of Utah College of Medicine, Salt Lake City, UT 84132

Tamas L. Frigyesi, MD [171]
EPID Research Institute, Lubbock, TX 79413

Gerald D. Frye, PhD [99]
Department of Psychiatry, University of North Carolina School of Medicine, Chapel Hill, NC 27514

Jacob M. Hiller, PhD [29]
Department of Psychiatry, New York University Medical Center, New York, NY 10016

Ryan J. Huxtable, PhD [53]
Department of Pharmacology, University of Arizona, Tucson, AZ 85724

The bold face numbers in brackets following each contributor's name indicate the opening page number of that author's paper.

Graham A.R. Johnston, PhD **[1]**
Department of Pharmacology, University of Sydney, Sydney, NSW, 2006, Australia

Alexander D. Kenny, PhD **[xiii]**
Department of Pharmacology and Therapeutics, Texas Tech University Health Sciences Center, Lubbock, TX 79430

K. Krnjević, MB, ChB, PhD **[127]**
Departments of Physiology and Anaesthesia Research, McGill University, Montreal, Quebec, Canada H3G 1Y6

I.M. Lang, PhD **[197]**
Department of Physiology, Texas Tech University Health Sciences Center, Lubbock, TX 79430

J.B. Lombardini, PhD **[171]**
Department of Pharmacology and Therapeutics, Texas Tech University Health Sciences Center, Lubbock, TX 79430

Richard B. Mailman, PhD **[99]**
Departments of Psychiatry and Pharmacology, University of North Carolina School of Medicine, Chapel Hill, NC 27514

Hugh McLennan, PhD **[19]**
Department of Physiology, University of British Columbia, Vancouver, British Columbia, Canada V6T 1W5

K. John Morrow, PhD **[151]**
Department of Biochemistry, Texas Tech University Health Sciences Center, Lubbock, TX 79430

Robert A. Mueller, PhD, MD **[99]**
Departments of Anesthesiology and Pharmacology, University of North Carolina School of Medicine, Chapel Hill, NC 27514

T. Celeste Napier, BS **[191]**
Department of Pharmacology and Therapeutics, Texas Tech University Health Sciences Center, Lubbock, TX 79430

Vernon J. Perez, PhD **[151]**
Department of Biochemistry, Texas Tech University Health Sciences Center, and Department of Psychology, Texas Tech University, Lubbock, TX 79409

James H. Pirch, PhD **[191]**
Department of Pharmacology and Therapeutics, Texas Tech University Health Sciences Center, Lubbock, TX 79430

Howard K. Strahlendorf, PhD **[161, 197, 217]**
Department of Medical and Surgical Neurology, Texas Tech University Health Sciences Center, Lubbock, TX 79430

Jean C. Strahlendorf, PhD **[161, 197, 217]**
Department of Physiology, Texas Tech University Health Sciences Center, Lubbock, TX 79430

Thomas D. Thomson, DVM, PhD **[181]**
Department of Pharmacology, University of Utah College of Medicine, Salt Lake City, UT 84132

Earl A. Zimmerman, MD **[117]**
Department of Neurology, College of Physicians and Surgeons, Columbia University, New York, NY 10032

Preface

The Third Tarbox Parkinson's Disease Symposium was held October 16–18, 1980, at the South Park Inn in Lubbock, Texas; these Proceedings represent the papers that were presented at this event. The Symposium was sponsored by the Tarbox Parkinson's Disease Institute and the Department of Pharmacology and Therapeutics of the Texas Tech University Health Sciences Center at Lubbock. The First Symposium took place in October 1976 and the Second Symposium was held in February 1978.

The Third Tarbox Parkinson's Disease Symposium brought together investigators actively engaged in research concerning the role of peptides and amino acids as neurotransmitters. The program, conceived and assembled by the Program Chairman, Dr. J. Barry Lombardini, was notable from several aspects. First, seven prominent investigators representing different disciplines were brought to Lubbock for the purpose of encouraging communication across these disciplines and diverse research areas. Secondly, several of the investigators, although known to each other by name and reputation, had never met on a prior occasion. Thirdly, part of the success of the Symposium may be judged by the fact that certain of the participants are already collaborating on projects that would not have developed without the contact made possible by the Symposium. The final verdict on the success of the event must await the outcome of the publication of this volume. Nevertheless, those in attendance could not fail but note the enthusiasm and excitement of the participants as they presented their findings. This in itself is testimony that the Symposium accomplished some degree of success.

The Tarbox Parkinson's Disease Institute was established in 1973 with funds appropriated by the State of Texas and is dedicated to research, clinical activities, and education in Parkinson's disease and related neurological disorders. The Institute is named after Mr. Elmer L. Tarbox, who served the Lubbock area as a Representative in the Texas Legislature, and is himself a victim of Parkinson's disease.

The Tarbox Fellowship Programs were instituted in 1977 and represent the research thrust of the Tarbox Institute. These include the Tarbox Postdoctoral Fellowships, the Tarbox Predoctoral Research Fellowships, and the Tarbox Medical Student Research Fellowships. The Institute also conducts a regularly scheduled Tarbox Clinic for Parkinson's disease at the Texas Tech University Health Sciences Center in Lubbock.

Thanks are especially due to the Faculty and Staff of the Department of Pharmacology and Therapeutics and to the Staff of the Office of Continuing Medical Education. The Symposium could not have succeeded without their help.

Alexander D. Kenny, PhD

Acknowledgments

The Third Tarbox Parkinson's Disease Symposium was supported in part by grants from The Upjohn Company and Sandoz, Inc.

The Role of Peptides and Amino Acids as Neurotransmitters, pages 1–17
©1981 Alan R. Liss, Inc., 150 Fifth Avenue, New York, NY 10011

GABA Receptors

Graham A.R. Johnston

The simple amino acid GABA, γ-aminobutanoic acid, is the major inhibitory transmitter in most regions of the brain being involved in at least one-third of all synaptic transmission. This statement is based on a wealth of evidence from neuropharmacological and neurochemical studies concerning the synthesis, storage, release, action, inactivation, and metabolism of GABA, which has been the major topic of numerous reviews [eg, Tapia, 1975; Ryall, 1975; Kelly and Beart, 1975; Curtis and Johnston, 1975; Johnston, 1978; Saelens and Vinick, 1978; Sytinsky et al, 1978; Andrews and Johnston, 1979a; Curtis, 1979; Enna and Maggi, 1979; Peck, 1980], books [eg, Davidson, 1976; Oja et al, 1977], and conferences [eg, Roberts et al, 1976; Fonnum, 1978; Krogsgaard-Larsen et al, 1979; Mandel and DeFeudis, 1979]. Abnormalities in GABA systems are a common feature of extrapyramidal disorders including Parkinson's disease [Marsden, 1978; Lloyd, 1980].

Studies on the neuronal uptake of GABA and on the localization of GAD (L-glutamate decarboxylase), the enzyme that catalyses the synthesis of GABA, indicate that presynaptic GABA terminals constitute between 30 and 45% of all presynaptic terminals in all brain regions studies thus far [Fonnum and Storm-Mathisen, 1978]. Most GABA neurones appear to be, in general, intrinsic to particular brain structures, ie, interneurones, though some are extrinsic, eg, Purkinje cells which project from the cerebellar cortex to Deiter's nucleus. Of particular interest to this Parkinson's disease symposium are the GABA neurones which project to the substantia nigra from the striatum and globus pallidus. As intrinsic neurones, GABA neurones are likely to be involved in most neuronal processing as befits the major inhibitory transmitter. GABA neurones appear to be used less often to transmit "inhibitory information" from one brain region to another.

It is pertinent to compare GABA to a more traditional transmitter such as dopamine. The levels of GABA and dopamine in certain regions of the human brain are shown in Table I; even in the caudate nucleus, one of the regions of

TABLE I. Levels of GABA and Dopamine in the Human Brain (nmole/g)

Region	GABA[a]	Dopamine[b]
Cerebral cortex	2100	0.8
Caudate nucleus	3000	37.0
Thalamus	2500	3.0
Substantia nigra	5300	2.5

[a]Perry et al [1971].
[b]Hornykiewicz [1966].

highest dopamine levels, the levels of GABA are two orders of magnitude higher than those of dopamine. Dopamine appears to be a specialized transmitter released by only a small percentage of neurones in most brain areas. Dopamine neurones appear to be mainly extrinsic, transmitting their information from one brain region to another as in the well-known strio-nigral pathway. They may be thought of as mainly specialized output neurones. Malfunction of dopamine neurones is widely considered to underly a variety of neurological disorders including Parkinson's disease. It may well be that malfunction of such specialized output neurones results from deficiencies in the network of processing neurones that control the output neurones. Thus, the malfunction of dopamine neurons may often represent the neurophysiological symptoms rather than the cause of a particular neurological disorder.

Since it is the major inhibitory transmitter, GABA is likely to be involved in most neurological disorders. We need to be able to develop much more specific therapeutic approaches to the treatment of abnormalities in GABA-mediated synaptic transmission than is the case with specialized transmitters such as dopamine. Recent advances in our understanding of basic aspects of GABA neuropharmacology indicate that the GABA system is extraordinarily complex compared to the systems associated with the more traditional transmitters. We are learning to appreciate that GABA has a rather different role to play in nervous system function since it is involved in such a large proportion of synaptic transmission in the brain. This review describes the complexities of GABA receptors which appear to exist in a number of domains in conjuction with ionophores and various modulator sites for drugs such as barbiturates and benzodiazepines.

HOW TO STUDY GABA RECEPTORS?

The activation of GABA receptors may be studied in vivo using electrophysiological or behavioral methods and in vitro using isolated tissue preparations. Recently, such studies have been supplemented by neurochemical investigations of GABA binding to a variety of membrane preparations. Such binding studies provide direct information on the affinity of the interactions between GABA and its binding sites, but only very indirect information on the relevance of these interactions.

Binding to GABA recognition sites on the receptors is only the first stage in producing the changes in ionic permeability that influence membrane potential and neuronal excitability. In very simplistic terms, the sequence of events in the hyperpolarization of a neurone by synaptically released GABA might be: (1) GABA binds to its recognition site on the receptor, (2) this binding produces a change in the conformation of the receptor, (3) the activated receptor produces a change in an associated ionophore, (4) the activated ionophore allows chloride ions to pass through the neuronal membrane, and (5) this flow of chloride ions hyperpolarizes the neurone making it less excitable. This sequence appears to be modulated by agents such as barbiturates, which appear to potentiate the coupling between receptors and ionophores. GABA antagonists appear to act on the recognition site (eg, bicuculline) and/or on the ionophore (eg, picrotoxin). Thus, GABA receptors might be considered as part of complex domains consisting of receptors, modulator sites, and ionophores, with all three influencing the final outcome of the initial activation of the GABA recognition site.

GABA can, in addition to hyperpolarizing neurones, have a depolarizing action. While GABA hyperpolarizes most neuronal cell bodies, it depolarizes some presynaptic terminals and nonmyelinated nerve fibers.

GABA Analogs

The assessment of GABA receptors requires extensive information on the properties of GABA analogs. There is an ever-increasing range of GABA analogs that is being studied. In the main, these are conformationally restricted analogs of GABA representing particular shapes of the flexible GABA molecule. As a result of free rotation around the single bonds in its carbon chain, the GABA molecule has considerable flexibility. This flexibility can be reduced in analogs of GABA by incorporation of unsaturation (as in trans-4-aminocrotonic acid), ring structures (as in cis-3-aminocyclopentane carboxylic acid), or both (as in muscimol) into the basic transmitter molecule. A systemic study of GABA analogs of restricted conformation has shown that GABA interacts in different shapes with various receptors, transport carriers, and enzymes [Johnston et al, 1978].

GABA does not pass the blood–brain barriers readily on systemic administration and considerable effort has been devoted to preparing relatively lipophilic derivatives of GABA that might be useful therapeutic agents [Bartholini and Lloyd, 1980]. Baclofen was developed for this purpose, although its therapeutic action in the treatment of spinal spasticity is more likely to be the result of inhibition of excitant amino acid release [Kato et al, 1980]. Muscimol, which is an isoxazole isolated from the mushroom *Amanita muscaria,* does produce central effects on systemic administration, but these could be due to metabolites of muscimol [Enna and Maggi, 1979]. Analogs of muscimol, such as THIP [Krogsgaard-Larsen et al, 1977], and other heterocyclic analogs of GABA, such as kojic amine [Atkinson et al, 1979], are being investigated for their therapeutic potential. SL 76-002 is a Schiff base of GABA amide developed as a systemically active GABA agonist [Lloyd et al, 1979], but its effects might also be due to metabolites. Esters of GABA with long-chain alcohols, eg, cetyl GABA, show some promise as therapeutic agents [Sytinsky et al, 1978; Frey and Loscher, 1980].

Many more GABA analogs are needed, though the chemistry of these structurally simple molecules is very difficult. It would seem logical to combine the above approaches and develop lipophilic derivatives of some of the conformationally restricted analogs of GABA.

Electrophysiological Studies

GABA antagonists. Many substances inhibit neuronal firing when administered extracellularly by electrophoresis from a microelectrode. In order to characterize such an inhibitory action as likely to be the result of activation of GABA receptors, it is necessary to study the effects of a variety of antagonists that are known from previous work to antagonize selectively the action of particular inhibitors; the circular arguments involved in such studies should be readily apparent! In general, the convulsant alkaloid bicuculline antagonizes the activation of GABA receptors, while the convulsant alkaloid strychnine does not. On this basis GABA agonists may be defined as bicuculline-sensitive, strychnine-insensitive neuronal inhibitors. This distinguishes them from glycine agonists, defined as strychnine-sensitive, bicuculline-insensitive neuronal inhibitors, and from substances such as dopamine and noradrenaline, whose inhibitory actions are insensitive to both alkaloids [Johnston, 1978]. This definition of GABA agonists represents a gross oversimplification, since there is evidence for the existence of bicuculline-insensitive GABA receptors and, furthermore, of a multiplicity of bicuculline-sensitive GABA receptors. The classification of GABA receptors seems likely to be even more complicated than the current situation with adrenoceptors, and more selective GABA antagonists are vital to any further progress.

Picrotoxin is widely used as a selective GABA antagonist (although it also antagonizes some actions of serotonin), but recent electrophysiological studies on cuneate slices have indicated that it acts at different sites to bicuculline [Simmonds, 1980a]. This is in agreement with binding studies that suggest picrotoxin interacts with ionphores rather than with GABA binding sites [Olsen et al, 1979]. Thus, picrotoxin might be more appropriately classified as an ionophore antagonist.

GABA agonists. The most extensive electophysiological studies of GABA agonists and antagonists in vivo have been made on spinal interneurones in anesthetized cats. These studies have included GABA analogs of restricted conformation resulting from unsaturation [Beart et al, 1971; Johnston et al, 1975], isoxazole [Krogsgaard-Larsen et al, 1975], tetrahydropyridine [Krogsgaard-Larsen et al, 1977], cyclopentane [Allan et al, 1980], or cyclopropane rings [Allan et al, 1980]. GABA receptors may occur as widely on nerve terminals as they do on nerve cell bodies [Pickles, 1979]. Presynaptic GABA receptors on primary afferent terminations in cat cord appear to be very similar to postsynaptic GABA receptors in this tissue [Curtis et al, 1980]. Only limited structure–activity studies have been reported on the action on cat spinal neurones of the GABA antagonist bicuculline and the ionophore antagonist picrotoxin [Curtis and Johnston, 1974; Johnston, 1976]. Though much more work needs to be done, particularly with respect to a more quantitative assessment of agonist and antagonist activity, a pattern has emerged indicating the range of "active conformations" of GABA agonists and antagonists at these feline, spinal postsynaptic receptors [Andrews and Johnston, 1979b].

Although less extensive in vivo electrophysiological studies have been made in other regions of the cat CNS and in other species, there is sufficient evidence available to suggest that there are both regional and species differences in GABA receptors as studied in vivo using the microelectrophoretic technique. Thus the relative potency of GABA and muscimol as bicuculline-sensitive, strychnine-insensitive neuronal inhibitors differs in the cat cortex and spinal cord [Curtis et al, 1971], while glycine appears to be able to activate bicuculline-sensitive GABA receptors in the rat cerebral cortex [Biscoe et al, 1972]. Structure–activity studies on GABA receptors in the neurohypophysis suggest some differences from spinal receptors [Mathison and Dreifuss, 1980].

Some important electrophysiological studies have been carried out on isolated tissues in vitro. In crayfish muscle, four types of GABA receptors have been characterized on the basis of desensitization experiments and antagonist effects of 3-guanidinopropionic acid, and these receptor types appear to be different from the major types described in the mammalian CNS [Dudel and Hatt, 1976].

Clear evidence for different classes of bicuculline-sensitive GABA receptors has been provided in studies comparing a range of GABA agonists in their ability to depress spontaneous synaptic activity in isolated immature rat hemisected spinal cord and to depolarize isolated spinal roots [Allan et al, 1980]. The relative depressant potencies were very similar to those found in cat spinal cord in vivo, whereas the relative depolarizing potencies were strongly correlated with data for the displacement of GABA bound to rat brain membranes. These experiments raise the important possibility that binding experiments are measuring nonsynaptic binding sites related to GABA receptors found on nerve fibers rather than binding sites associated with synaptic receptors. The nonsynaptic receptors present on spinal roots are probably identical to the receptors which are found on unmyelinated peripheral nerve fibers [Brown and Marsh, 1978]. It seems likely that both synaptic and nonsynaptic types of GABA receptors occur in the CNS, since GABA exhibits biphasic responses on hippocampal neurones producing depolarization and hyperpolarization at different sites [Alger and Nicoll, 1979].

Bicuculline-insensitive neuronal inhibition by certain GABA analogs in folded conformations has been described on cat spinal neurones. These analogs include baclofen [Curtis et al, 1974], cis-4-aminocrotonic acid [Johnston et al, 1975], trans-2- and trans-3-aminocyclohexane carboxylic acid [Johnston, 1976], and cis-2-(aminomethyl)-cyclopropane carboxylic acid [Allan et al, 1980]. Baclofen has been recently proposed as a selective agonist for a novel bicuculline-insensitive class of receptors which modulate monoamine release [Bowery et al, 1980]. It seems likely that bicuculline-insensitive GABA receptors may be widely distributed in the CNS.

GABA modulators. Barbiturates and benzodiazepines potentiate GABA-mediated synaptic transmission in many areas of the CNS [eg, Ransom and Barker, 1976; Polc and Haefely, 1976; Nicoll, 1978; MacDonald and Barker, 1978]. The effects of these drugs are not the result of a decreased rate of inactivation of GABA, since these drugs inhibit the active uptake of GABA only at relatively high concentrations. They appear to influence GABA receptor–ionophore complexes such that GABA responses are potentiated.

There is electrophysiological evidence that barbituarates and benzodiazepines act in different ways to potentiate GABA responses. It is often found that barbiturates, but not benzodiazepines, potentiate neuronal effects of locally administered GABA, while both classes of drugs potentiate GABA-mediated synaptic transmission. Many studies suggest that benzodiazepines act presynaptically, increasing GABA release [eg, Polc et al, 1974; Curtis et al, 1976]. Barbiturates appear to act on GABA release and also postsynaptically at sites closely related to bicuculline antagonism sometimes exhibiting GABA agonist activity [Ransom and Barker, 1976; Bowery and Dray, 1978; Nicoll, 1978; Lodge and Curtis, 1978]. Benzodiazepines may act on the picrotoxin-sensitive ionophore sites in potentiating GABA responses in cuneate slices [Simmonds, 1980b].

Behavioral Studies

Various animal models have been developed for evaluating drug effect on GABA systems in vivo [eg, Scheel-Kruger et al, 1979; Waddington and Cross, 1979]. For example, unilateral injection of the GABA agonist muscimol into the zona reticulata of the substantia nigra induces strong contralateral turning in rats, while injection of the GABA antagonists picrotoxin and bicuculline produces ipsilateral turning. Barbiturates and benzodiazepines potentiate many of the behavioral effects of GABA agonists.

The results of such behavioral experiments with GABA-related substances are usually more difficult to interpret than those of similar experiments using substances related to dopamine reflecting the different roles of the two transmitters. Intranigral GABA potentiates dopamine-dependent stereotyped behavior but decreases dopamine-dependent locomotor activity, indicating the connections between GABA, dopamine, and any observed behavior are very complicated.

An increased response to intranigral injection of muscimol following lesion of the strio-nigral tract and degeneration of GABA neurones in the substantia nigra provides evidence for denervation supersensitivity of certain GABA receptors [Waddington and Cross, 1978].

Baclofen and muscimol appear to have different behavioral effects [Delini-Stula, 1979], eg, on intranigral injection [Waddington and Cross, 1979]. Baclofen may have a partial agonist action at certain nigral GABA receptors, perhaps those related to dopamine release as suggested by the work of Bowery et al [1980].

Binding Studies

The first binding studies relevant to GABA receptors were carried out by Peck et al [1973] on the bicuculline-sensitive binding of radioactive GABA to synaptosomal membranes from rat brain in the absence of sodium ions and in the presence of chlorpromazine (to prevent binding to uptake carriers). Subsequent studies, particularly by the groups of Enna [eg, Enna et al, 1979] and Olsen [eg, Olsen et al, 1979], have provided a huge amount of information on the binding of GABA and related compounds, eg, muscimol and dihydropicrotoxinin, to a variety of membrane preparations.

Most GABA binding is associated with synaptic membranes, and structure–activity studies show a reasonable correlation between the ability of analogs to displace bound GABA and with their electrophysiological potencies as GABA receptor agonists. Lesioning and tissue culture experiments suggest that most GABA binding is associated with neurones.

The kinetics of GABA binding to brain membranes are usually biphasic, being analysed in terms of "high-" and "low-"affinity binding sites. These are likely to be relatively independent, separate sites, since their relative proportions show regional variation [Enna et al, 1979], and they differ in heat inactivation rates and sensitivity to changes in pH, temperature, and ionic environment [Olsen, 1980]. The slope of the Hill plots for each of the sites equals one indicating no cooperativity between the sites [Horng and Wong, 1979]. Little is known about the probably functional roles of these separate sites. One possibility is that the low-affinity sites correspond to autoreceptors that may modulate GABA release from presynaptic terminals and respond to high concentrations of GABA in the synaptic cleft that would otherwise saturate high-affinity binding sites. Activation of low-affinity presynaptic sites might reduce GABA release and serve to maintain optimum synaptic concentrations of the transmitter. Evidence for such GABA autoreceptors comes from in vitro release experiments in which GABA release can be potentiated by bicuculline [Johnston and Mitchell, 1971] and inhibited by muscimol [Snodgrass, 1978].

In contrast to the many studies on agonist binding, relatively few studies have been carried out with antagonists. Bicuculline methiodide appears to bind to only a single population of GABA-sensitive sites with relatively low affinity; such binding is enhanced by various anions which decrease GABA binding and this has led to the proposal of interconvertible agonist and antagonist binding sites [Mohler and Okada, 1978]. Dihydropicrotoxinin binds to GABA-insensitive sites which appear to be associated with GABA-activated chloride ionophores [Olsen et al, 1979]. A number of convulsant drugs including barbiturates, bicyclophosphates, and bicyclocarboxylate ortho-esters displace dihydropicrotoxinin, but not GABA, from binding sites on rat brain membranes. There are far fewer dihydropicrotoxinin binding sites than GABA binding sites on rat brain membranes, suggesting that not all GABA binding sites are associated with ionophores, but this may be misleading due to the complications of endogenous inhibitors of the binding of GABA, as discussed later, and of the binding of dihydropicrotoxinin [Olsen and Leeb-Lundberg, 1980].

Electrophysiological studies provide evidence for different classes of GABA receptors. Thus, binding might reflect the interaction of GABA and recognition sites on postsynaptic, presynaptic, and nonsynaptic receptors that might be expected to show different binding characteristics. In addition, binding may occur to the active sites on enzymes that metabolize GABA and to the recognition sites on carriers that transport GABA into glia and neurones. Binding studies must rely heavily on what is known of the pharmacology and biochemistry of these interactions from other types of experiments. In general "receptor" binding studies are carried out in the absence of sodium ions, eliminating sodium-dependent binding processes that are associated with transport carriers, and use is made of agents considered to show some selectivity for receptor-related interactions. Muscimol is widely used as a selective GABA agonist, but there are problems associated with this in that not all muscimol binding sites are related to GABA binding sites, and not all GABA binding sites are related to muscimol binding sites [Johnston et al, 1978; DeFeudis, 1980]. Furthermore, bicuculline sensitivity, as discussed above, is not a characteristic of all GABA receptors.

Structure–activity studies on GABA binding show that different membrane preparations may favor different classes of GABA binding sites [Krogsgaard-Larsen et al, 1979]. This aspect should be investigated more extensively in view of the recent study which described a strong correlation between potency in displacing GABA bound to rat brain membranes prepared by a freeze–thaw regime, and potency in activating GABA nonsynaptic receptors on nerve fibers [Allan et al, 1980]. Thus, the majority of GABA binding studied in many experiments may reflect interaction with nonsynaptic receptors, rather than with pre- or postsynaptic receptors. More selective agents for the various types of GABA receptors would help to sort out these problems.

GABA binding studies have provided much information that would not have been obtained easily in any other way. Most important have been the discovery of endogenous inhibitors of GABA binding other than GABA itself and a better understanding of the ways in which barbiturates and benzodiazepines might modulate the activity of GABA receptors and associated ionophores.

GABARINs: Endogenous Inhibitors

The kinetics of the binding of GABA and muscimol of rat brain membranes are dependent on the methods used to prepare the membranes, detergent extraction enhancing both the affinity of binding and the number of binding sites. The exposure of these latent GABA binding sites is due to the removal of substances that inhibit GABA binding and which are normally incorporated into the membranes. The detergent treatment (and also freeze–thaw regimes) renders these inhibitors soluble, and their presence in the supernatant washes of such membrane preparations is readily demonstrated. The existence of these inhibitors was first described at the meeting on "Amino Acids as Chemical Transmitters" in Norway in 1977 [Johnston and Kennedy, 1978; Olsen et al, 1978]. A variety of such inhibitors appears to exist, and I have called them collectively GABARINs (*GABA R*eceptor *IN*hibitor*s*).

Removal of GABARINs could underlie phenomena such as denervation supersensitivity. Increased apparent GABA binding density has been noted in the substantia nigra following striatal lesions accompanied by an increased behavioral response to intranigral muscimol injection as noted above [Waddington and Cross, 1978]. GABARINs might be involved in many aspects of the function of this major inhibitory transmitter. They may be involved in desensitization and supersensitivity phenomena. They may play an important role in synaptic development and in synaptic changes involved in learning and memory. Many drugs are likely to affect GABARINs, causing altered functioning of GABA receptor complexes.

Occluded GABA. Some, but not all, of the GABARIN activity in detergent extracts of rat brain membranes can be attributed to GABA originally occluded in the membranes [Johnston and Kennedy, 1978; Olsen, et al, 1978]. Such GABA "in the occult form" [Elliott and Van Gelder, 1958] may represent a storage or inactivated form of GABA. It is in nonexchangeable, osmotically insensitive compartments and influences both the apparent availability and affinity of GABA binding sites.

Occluded GABA could be the basis of the apparent desensitization of GABA responses seen in vivo, GABA combining with the receptor and both being buried in the membrane ("cryptic receptors").

Occluded GABA should not be confused with mitochondrial and other loosely bound GABA liberated during the early stages of membrane preparation by homogenization and osmotic shock. This loosely bound GABA certainly inhibits the binding of radioactive GABA, unless it is removed by thorough washing. Treatment of a thoroughly washed membrane preparation with detergent liberates the occluded GABA; this tightly bound GABA does not exchange with radioactive GABA added in the initial homogenization and thus does not represent redistributed GABA [Johnston, unpublished].

Phosphatidylethanolamine. Phospholipids are recognized to be important in the interactions of many transmitters and hormones with their receptors [Loh and Law, 1980]. In view of the structural similarities between GABA and the polar head group of phosphatidylethanolamine, Watkins [1965] proposed that GABA might induce membrane changes by acting on a phosphatidylethanolamine-receptor–protein complex. This is now appreciable evidence to support a role for phosphatidylethanolamine in GABA receptor function. Treatment of crude synaptic membranes with phospholipase C increases the availability and affinity of GABA binding sites [Giambalvo and Rosenberg, 1976], as does enzymic base exchange with ethanolamine [Johnston and Kennedy, unpublished], while phosphatidylethanolamine inhibits GABA binding to detergent-extracted brain membranes [Johnston and Kennedy, 1978]. Phospholipids are abundant in detergent extracts of brain membranes, and it seems very likely that derivatives of phosphatidylethanolamine, perhaps with different fatty acid residues, can act as endogenous inhibitors of GABA binding [Johnston and Kennedy, 1978].

There is evidence that a deficiency of phospholipid inhibitor may be responsible for the increased number of GABA binding sites found in some brain areas of patients with Huntington's disease [Lloyd and Davidson, 1979]. Increased brain levels of glycerophosphoethanolamine are found in Huntington's disease, and it is possibly that abnormal phospholipid metabolism is linked to the neuropathogenesis of many neurological diseases.

Peptides. An acidic peptide, "GABA-modulin," of molecular mass 15,000 daltons, has been reported to inhibit the binding of GABA to sites on brain membranes [Toffano et al, 1978]. GABA-modulin also inhibits benzodiazepine binding, but is some 30 times weaker than as an inhibitor of GABA binding [Guidotti et al, 1978]. Although much has been written about GABA-modulin by the group that discovered it, widespread support for the function of this peptide unfortunately has not been forthcoming. GABA-modulin may be very susceptible to breakdown by endogenous proteases and a reliable preparation of this peptide would enable it to be studied more extensively.

Peptides of relatively low molecular weight which inhibit GABA binding can be separated from detergent extracts of brain membranes [Johnston and Kennedy,

1979 and unpublished]. Unlike GABA-modulin they do not inhibit benzodiazepine binding. These could be breakdown products of GABA-modulin, or they could have a similar relationship to the larger peptides as do enkephalins to endorphins.

GRIF and purines. Yoneda and Kuriyama [1980] have described a low-molecular weight nonpeptide inhibitor (GRIF) of muscimol binding in extracts of rat brain membranes. This *G*ABA *R*eceptor *I*nhibitory *F*actor (GRIF) can be separated readily from GABA by cation exchange chromatography. It is stable to heat, trypsin, collagenase, peroxidase, and phospholipase C treatment and has a molecular mass of less that 500 daltons based on gel and ultrafiltration studies.

GRIF could be a purine or related substance. Several purines, including adenosine, hypoxanthine, and inosine, inhibit GABA binding [Ticku and Burch, 1980] and are thus possible endogenous GABARINs. Purine inhibition of GABA binding is noncompetivive in nature, while purine inhibition of benzodiazepine binding is competitive and shows a different structural specificity. Purines are relatively weak inhibitors (IC50 values ca. 0.5 mM) of GABA binding, but it is not impossible that they occur in sufficiently high concentrations to influence the activity of GABA receptors. Of course, purines are being investigated in their own right as neurotransmitters, and modulation of the GABA system may be an important function of certain purines.

Enhancement of GABA Binding

The two groups of drugs that are known to modulate GABA action, the benzodiazepines and barbiturates, can be shown to stimulate GABA binding under appropriate circumstances. GABA receptor–ionophore complexes appear to be composed on a number of interacting domains. Depending on the membrane preparation, these domains may be studied in relative isolation or in varying degrees of interaction. In order to observe enhancement of GABA binding by barbiturates and benzodiazepines, it is necessary to use a membrane preparation that preserves the interactions between the GABA binding sites and the modulator binding sites [Johnston and Willow, 1981]. It is clear that barbiturates and benzodiazepines act in different ways to enhance GABA binding, consistent with electrophysiological observations.

Chronic administration of phenobarbitone or diazepam to rats leads to a decrease in the number of high-affinity GABA binding sites in the striatum [Mohler et al, 1978].

Benzodiazepines. There have been many papers on the action of GABA and various analogs in enhancing the binding of benzodiazepine. This action has a unique structural specificity that suggests that a particular population of GABA binding sites is involved, and certain analogs, such as THIP, that are potent GABA agonists may in vivo situations exhibit only partial agonist properties in stimulating benzodiazepine binding [Karobath et al, 1979]. Low-affinity GABA binding sites appear to be responsible for the enhancement of benzodiazepine binding [Stone and Topham, 1980]. Supersensitivity to the enhancement of benzodiazepine binding by GABA develops after striatonigral lesion [Shibuya et al, 1980], and barbiturates potentiate the effects of GABA on benzodiazepine binding [Skolnick et al, 1980]. Potentiation of benzodiazepine binding by GABA is not paticularly dependent on the membrane preparation and has even been described with soluble benzodiazepine binding preparations [Gavis and Snyder, 1979]. This enhancement is, thus, unlikely to be due to the displacement of endogenous inhibitors of benzodiazepine binding.

It is more difficult to demonstrate the enhancement of GABA binding by benzodiazepines. This needs a relatively intact membrane preparation and may well be due to the displacement of endogenous inhibitors of GABA binding [Guidotti et al, 1978; Johnston and Willow, 1980]. Thus, there appear to be two different phenomena involved in that the enhancement of GABA binding by benzodiazepines is different from the enhancement of benzodiazepine binding by GABA.

Barbiturates. A range of barbiturates enhances the binding of GABA to relatively intact membrane preparations from rat brain [Johnston and Willow, 1981 and unpublished]. The enhancement appears to be a competitive effect on the low-affinity binding site, and it can be blocked by picrotoxin which, however, has no effect on GABA binding per se to these membrane preparations.

Barbiturates have other actions on other transmitter systems, influencing transmitter release probably via effects on presynaptic calcium–ATPase activity [Willow and Johnston, 1979]. This may explain why babiturates are less selective drugs than benzodiazepine.

Classification of GABA Receptors

There are many different types of GABA receptors and the following classification is proposed in order to aid further work, particularly with respect to the development of more selective agents with which to investigate the functions of this important transmitter.

Anatomical:	A—Postsynaptic receptors B—Presynaptic receptors C—Nonsynaptic receptors
Physiological:	H—Hyperpolarizing receptors D—Depolarizing receptors
Pharmacological:	1—Bicuculline sensitive 2—Bicuculline insensitive 3—Benzodiazepine sensitive 4—Benzodiazepine insensitive 5—Barbiturate sensitive 6—Barbiturate insensitive

On such a classification the classical GABA receptors on spinal motoneurones of the cat might be classified as AH-145, being postsynaptic hyperpolarizing receptors sensitive to bicuculline and barbiturates, but not benzodiazepines.

Obviously, this is a very complicated classification and one that I expect will get more complicated with the development of more selective agents!

CONCLUSION

This review has emphasized that there is a great deal still to be learned about GABA as a neurotransmitter. GABA systems appear to be richly complex. The very complexity of the GABA receptor–modulator–ionophore domains gives hope for the development of very selective therapeutic agents which will be important in the treatment of many neurological disorders.

ACKNOWLEDGMENTS

The author is grateful to his colleagues R.D. Allan, J.H. Skerritt, and M. Willow for their helpful advice and collaboration.

DEDICATION

This article is dedicated to Charles and Eileen Shoppee. Charles Shoppee was Professor of Organic Chemistry at the University of Sydney 1956–1969, where he guided me in my first research work, and subsequently he spent several years as a Visiting Professor at Texas Tech University. His wife, Eileen, died of Parkinson's disease in August 1979.

REFERENCES

1. Alger BE, Nicoll RA: GABA-mediated biphasic inhibitory responses in hippocampus. Nature (London) 218:315–317, 1979.
2. Allan RD, Curtis DR, Headley PM, Johnston GAR, Kennedy SME, Lodge D, Twitchin B: Cyclobutane analogues of GABA. Neurochem Res 5:393–400, 1980.
3. Allan RD, Curtis DR, Headley PM, Johnston GAR, Lodge D, Twitchin B: The synthesis and activity of cis- and trans-2-(aminomethyl)-cyclopropanecarboxylic acid as conformationally restricted analogues of GABA. J Neurochem 34:652–654, 1980.
4. Allan RD, Evans RH, Johnston GAR: GABA agonists: An in vitro comparison between depression of spinal synaptic activity and depolarization of spinal root fibers in the rat. Brit J Pharmac 70:609–615, 1980.
5. Andrews PR, Johnston GAR: GABA agonists and antagonists. Biochem Pharmacacol 28:2697–2702, 1979a.
6. Andrews PR, Johnston GAR: Conformational analysis of muscimol, a GABA agonist. J Theoret Biol 79: 263–273, 1979b.
7. Atkinson JG, Girard Y, Rokach J, Rooney CS, McFarlane CS, Rackham A, Share NN: Kojic amine—A novel gamma-aminobutyric acid analogue. J Med Chem 22:99–106, 1979.
8. Bartholini G, Lloyd KG: The potential of GABA-mimetics in the therapy of extrapyramidal disorders. J Neural Trans (Suppl) 16:229–238, 1980.
9. Beart PM, Curtis DR, Johnston GAR: 4-Aminotetrolic acid: A new conformationally-restricted analogue of gamma-aminobutyric acid. Nature New Biol 234:80–81, 1971.

10. Biscoe TJ, Duggan AW, Lodge D: Antagonism between bicuculline, strychnine and picrotoxin and depressant amino acids in the rat nervous system. Comp Gen Pharmacol 3:423–433, 1972.
11. Bowery NG, Dray A: Reversal of the action of amino acid antagonists by barbiturates and other hypnotic drugs. Brit J Pharmacol 63:197–215, 1978.
12. Bowery NG, Hill DR, Hudson AL, Doble A, Middlemiss DN, Shaw J, Turnbull M: (-)-Baclofen decreases neurotransmitter release in the mammalian CNS by an action at a novel GABA receptor. Nature (London) 283:92–94, 1980.
13. Brown DA, Marsh S: Axonal GABA-receptors in mammalian peripheral nerve trunks. Brain Res 156:187–191, 1978.
14. Curtis DR: Gabergic transmission in the mammalian central nervous system. In Krogsgaard-Larsen P, Scheel-Kruger J, Kofod H (eds): "GABA-Neurotransmitters." Copenhagen: Munksgaard, 1979, pp 17–27.
15. Curtis DR, Borstein JC, Lodge D: In vivo analysis of GABA receptors on primary afferent terminations in the cat. Brain Res 194:255–258, 1980.
16. Curtis DR, Duggan AW, Felix D, Johnston GAR, McLennan H: Antagonism between bicuculline and GABA in the cat brain. Brain Res 70:493–499, 1971.
17. Curtis DR, Game CJA, Johnston GAR, McCulloch RM: Central effects of beta-(p-chlorophenyl)-gamma-aminobutyric acid. Brain Res 70:493–499, 1974.
18. Curtis DR, Johnston GAR: Convulsant alkaloids. In Simpson LL, Curtis DR (eds): "Neuropoisons." Vol 2, New York: Plenum, 1974, pp 207–248.
19. Curtis DR, Johnston GAR: Amino acid transmitters in the mammalian mammalian central nervous system. Rev Physiol 69:97–188, 1975.
20. Curtis DR, Lodge D, Johnston GAR, Brand SJ: Central actions of benzodiazepines. Brain Res 118:344–347, 1976.
21. Davidson N: "Neurotransmitter Amino Acids." London: Academic Press, 1976.
22. DeFeudis FV: Binding studies with muscimol; relation to synaptic gamma-aminobutyrate receptors. Neuroscience 5:675–688, 1980.
23. Delini-Stula A: Differential effects of baclofen and muscimol on behavioral responses implicating GABA-ergic transmission. In Krogsgaard-Larsen P, Scheel-Kruger J, Kofod H (eds) "GABA-Neurotransmitters." Copenhagen: Munksgaard, 1979, pp 482–499.
24. Dudel J, Hatt H: Four types of GABA receptors in crayfish leg muscles characterized by desensitization and specific antagonist. Pflugers Arch 364:217–222, 1976.
25. Elliott KAC, Van Gelder NM: Occlusion and metabolism of gamma-aminobutyric acid by brain tissue. J Neurochem 3:28–40, 1958.
26. Enna SJ, Ferkany JW, Krogsgaard-Larsen P: Pharmacological characteristics of GABA receptors in different brain regions. In Krogsgaard-Larsen P, Scheel-Kruger J, Kofod H (eds) "GABA-Neurotransmitters." Copenhagen: Munksgaard, 1979, pp 191–200.
27. Enna SJ, Maggi A: Biochemical pharmacology of GABAergic agonists. Life Sci 19:1727–1738, 1979.
28. Fonnum F: "Amino Acids As Chemical Transmitters." New York: Plenum, 1978.
29. Fonnum F, Storm-Mathisen J: Localization of GABAergic neurons in the CNS. "Handbook of Psychopharmacology." Vol 9, New York: Plenum, 1978, pp 357–401.
30. Frey HH, Loscher W: Cetyl GABA: Effect on convulsant thresholds in mice and acute toxicity. Neuropharmacology 19:217–220, 1980.
31. Gavish M, Snyder SH: Soluble benzodiazepine receptors: GABAergic regulation. Life Sci 26:579–582, 1979.
32. Giambalvo C, Rosenberg P: The effect of phospholipase and proteases on the binding of gamma-aminobutyric acid to junctional complexes of rat cerebellum. Biochim Biophys Acta 436:741–756, 1976.

33. Guidotti A, Toffano G, Costa E: An endogenous protein modulates the affinity of GABA and benzodiazepine receptors in rat brain. Nature (London) 275:553–555, 1978.
34. Horng JS, Wong DT: Gamma-aminobutyric acid receptors in cerebellar membranes of rat brain after a treatment with Triton X-100. J Neurochem 32:1379–1386, 1979.
35. Hornykiewicz O: Dopamine (3-hydroxytyramine) and brain function. Pharmacol Rev 18:925–964, 1966.
36. Johnston GAR: Physiologic pharmacology of GABA and its antagonists in the vertebrate nervous system. In Roberts E, Chase TN, Tower DB (eds): "GABA in Nervous System Function." New York: Raven Press, 1976, pp 395–411.
37. Johnston GAR: Neuropharmacology of amino acid inhibitory transmitters. Ann Rev Pharmacol 18:269–289, 1978.
38. Johnston GAR, Allan RD, Andrews PR, Kennedy SME, Twitchin B: Stereospecific actions of GABA analogues. In Simon P (ed): "Advances in Pharmacology and Therapeutics." Vol 2, Oxford: Pergamon Press, 1979, pp 11–18.
39. Johnston GAR, Allan RD, Kennedy SME, Twitchin B: Systematic study of GABA analogues of restricted conformation. In Krogsgaard-Larsen P, Scheel-Kruger J, Kofod H (eds): "GABA-Neurotransmitters." Copenhagen: Munksgaard, 1979, pp 149–164.
40. Johnston GAR, Curtis DR, Beart PM, Game CJA, McCulloch RM, Twitchin B: Cis- and trans-4-aminocrotonic acid as GABA analogues of restricted conformation. J Neurochem 24:157–160, 1975.
41. Johnston GAR, Hailstone MH, Freeman CG: Baclofen: Stereoselective inhibition of excitant amino acid release. J Pharmacol Pharmac 32:230–231, 1980.
42. Johnston GAR, Kennedy SME: GABA receptors and phospholipids. In Fonnum F (ed): "Amino Acids as Chemical Transmitters." New York: Plenum, 1978, pp 507–516.
43. Johnston GAR, Kennedy SME: Endogenous inhibitors of GABA binding to rat brain membranes. Clin Exp Pharmac Physiol 6:686–687, 1979.
44. Johnston GAR, Mitchell JF: The effect of bicuculline, metrazol, picrotoxin and strychnine on the release of [3H]GABA from rat brain slices. J Neurochem 18:2441–2446, 1971.
45. Johnston GAR, Willow M: Barbiturates and GABA receptors. In Costa E, Di Chiara G, Gessa GL (eds): "GABA and Benzodiazepines." New York: Raven Press, 1981, pp 191–198.
46. Karobath M, Placheta P, Lippitsch M, Krogsgaard-Larsen P: Is stimulation of benzodiazepine receptor binding mediated by a novel GABA receptor? Nature (London) 278:748–749, 1979.
47. Kato M, Waldeman U, Murakami S: Effects of baclofen on spinal neurones of cats. Neuropharmacology 17:872–833, 1978.
48. Kelly JS, Beart PM: Amino acid receptors in CNS. II. GABA in supraspinal regions. In Iversen LL, Iversen SD, Snyder SH (eds): "Handbook of Psychopharmacology." Vol 4, New York: Plenum, 1975, pp. 129–209.
49. Krogsgaard-Larsen P, Hjeds, Curtis DR, Lodge D, Johnston GAR: Dihydromuscimol, thiomuscimol and related heterocyclic compounds as GABA analogues. J Neurochem 32:1717–1724, 1979.
50. Krogsgaard-Larsen P, Johnston GAR, Curtis DR, Game CJA, McCulloch RM: Structure and biological activity of a series of conformationally restricted analogues of GABA. J Neurochem 25:803–809, 1975.
51. Krogsgaard-Larsen P, Johnston GAR, Lodge D, Curtis DR: A new class of GABA agonist. Nature (London) 268:53–55, 1977.
52. Krogsgaard-Larsen P, Scheel-Kruger J, Kofod H: "GABA-Neurotransmitters." Copenhagen: Munksgaard, 1979.
53. Lloyd KG: The neuropathology of GABA neurons in extrapyramidal disorders. J Neural Trans (Suppl) 16:217–227, 1980.

54. Lloyd KG, Davidson L: [3H]GABA binding in brains from Huntington's chorea patients: Altered regulation by phospholipids? Science 205:1147–1149, 1979.
55. Lloyd KG, Worms P, Deportere H, Bartholini G: Pharmacological profile of SL 76002, a new GABA-mimetic drug. In Krogsgaard-Larsen P, Scheel-Kurger J, Kofod H (eds): "GABA-neurotransmitters." Copenhagen: Munksgaard, 1979, pp 308–325.
56. Lodge D, Curtis DR: Time course of GABA and glycine actions on cat spinal neurones: Effect of pentobarbitone. Neurosci Lett 8:125–129, 1978.
57. Loh HH, Law PY: The role of membrane lipids in receptor mechansims. Ann Rev Pharmacol 20:201–234, 1980.
58. MacDonald R, Barker JL: Benzodiazepines specifically modulate GABA-mediated postsynaptic inhibition in cultured mammalian neurones. Nature (London) 271:563–564, 1978.
59. Mandel P, DeFeudis FV: GABA Biochemistry and CNS Functions. New York: Plenum, 1979.
60. Marsden CD: GABA in relation to extrapyramidal diseases, with particular relevance to animal models. In Krogsgaard-Larsen P, Scheel-Kruger J, Kofod H (eds): "GABA-Neurotransmitters." Copenhagen: Munksgaard, 1979, pp 295–307.
61. Mathison RD, Dreifuss JJ: Structure-activity relationships of a neurohypophysial GABA receptor. Brain Res 187:476–480, 1980.
62. Mohler H, Okada T: Properties of gamma-aminobutyric acid receptor binding with (+)-[3H]bicuculline methiodide in rat cerebellum. Molec Pharmacol 14:256–265, 1978.
63. Mohler H, Okada T, Enna SJ: Benzodiazepine and neurotransmitter receptor binding in rat brain after chronic administration of diazepam or phenobarbital. Brain Res 156:391–394, 1978.
64. Nicoll R: Selective actions of barbiturates on synaptic transmission. In Lipton MA, Dimascio A, Killam KF (eds): "Psychopharmacology: A Generation of Progress." New York: Raven Press, 1978, pp 1337–1348.
65. Oja SS, Kontro P, Lähdesmäki P: Amino Acids as Inhibitory Neurotransmitters. "Progress in Pharmacology." Vol 1, No. 3, Stuttgart: Gustav Fischer, 1977.
66. Olsen RW: Biochemistry of GABA receptor proteins in mammalian brain. In Yamamura HI, Olsen RW, Usdin E (eds): "Psychopharmacology and Biochemistry of Neurotransmitter Receptors." New York: Elsevier, 1980, in press.
67. Olsen RW, Greenlee D, Van Ness P, Ticku MK: Studies on the gamma-aminobutyric acid receptor/ionophore proteins in mammalian brain. In Fonnum F (ed): "Amino Acids as Chemical Transmitters." New York: Plenum, 1978, pp 467–486.
68. Olsen RW, Leeb-Lundberg FL: Endogenous inhibitors of picrotoxin-convulsant binding sites in rat brain. Eur J Pharmacol 65:101–104, 1980.
69. Olsen RW, Ticku MK, Greenlee D, Van Ness P: GABA receptor and ionophore binding sites: interaction with various drugs. In Krogsgaard-Larsen P, Scheel-Kruger J, Kofod H (eds): "GABA-Neurotransmitters." Copenhagen: Munksgaard, 1979, pp 165–178.
70. Peck EJ: Receptors for amino acids. Ann Rev Physiol 42:615–627, 1980.
71. Peck EJ Schaeffer JM, Clark JH: Gamma-aminobutyric acid, bicuculline, and post-synaptic binding sites. Biochem Biophys Res Commun 52:394–400, 1973.
72. Perry TL, Berry K, Hansen S, Diamond S, Mok C: Regional distribution of amino acids in human brain obtained at autopsy. J Neurochem 18:513–519, 1971.
73. Pickles HG: Presynaptic gamma-aminobutyric acid responses in the olfactory cortex. Br J Pharmacol 65:223–228, 1979.
74. Polc P, Haefely W: Effects of two benzodiazepines, phenobarbitone and baclofen on synaptic transmission in the cat cuneate nucleus. Nauyn-Schmideberg's Arch Pharmacol 294:121–131, 1976.
75. Polc P, Mohler H, Haefely W: The effect of diazepam on spinal cord activities: Possible sites and mechanisms of action. Naunyn-Schmiedeberg's Arch Phamacol 284:319–337, 1974.

76. Ransom BR, Barker JL: Pentobarbital selectively enhances GABA-mediated post-synaptic inhibition in tissue cultured mouse spinal neurones. Brain Res 114:530–535, 1976.
77. Roberts E, Chase TN, Tower DB: "GABA in Nervous System Function." New York: Raven Press, 1976.
78. Ryall RW: Amino acid receptors in CNS. I. GABA and glycine in spinal cord. In Iversen LL, Iversen SD, Snyder SH (eds): "Handbook of Psychopharmacology." Vol 4, New York: Plenum, 1975, pp 83–128.
79. Saelens JK, Vinick FJ: Agents affecting GABA in the CNS. In Clarke FH (ed): "Annual Reports in Medicinal Chemistry." Vol 13, New York: Academic Press, 1978, pp 31–40.
80. Scheel-Kruger J, Arnt J, Braestrup C, Christensen AV, Magelund G: Development of new animal models for GABAergic actions using muscimol as a tool. In Krogsgaard-Larsen P, Scheel-Kruger J, Kofod H (eds): "GABA-Neurotransmitters." Copenhagen: Munksgaard, 1979, pp 447–464.
81. Shibuya H, Gale K, Pert CB: Supersensitivity to GABA's effect on benzodiazepine receptors develops after striatonigral lesions. Eur J Pharmacol 62:243–244, 1980.
82. Skolnick P, Paul SM, Barker JL: Pentobarbital potentiates GABA-enhanced [3H]-diazepam binding to benzodiazepine receptors. Eur J Pharmacol 65:125–127, 1980.
83. Simmonds MA: A site for the potentiation of GABA-mediated responses by benzodiazepines. Nature (London) 284:558–560, 1980a.
84. Simmonds MA: Evidence that bicuculline and picrotoxin act at separate sites to antagonize gamma-aminobutyric acid and its derivatives. Neuropharmacology 19:39–45, 1980b.
85. Snodgrass SR: Use of 3H-muscimol for GABA receptor studies. Nature (London) 273:392–394, 1978.
86. Stone MA, Topham LD: Does binding to the high and low affinity GABA binding sites binding to benzodiazepine receptor linked and non-benzodiazepine receptor linked GABA receptors? Br J Pharmacol 69:326P–327P, 1980.
87. Sytinsky IA, Soldatenkov AT, Lajtha A: Neurochemical basis of the therapeutic effect of gamma-aminobutyric acid and its derivatives. Prog Neurobiol 10:89–133, 1978.
88. Tapia R: Biochemical pharmacology of GABA in CNS. In Iversen LL, Iversen SD, Snyder SH (eds): "Handbook of Psychopharmacology." Vol 4, New York: Plenum, 1975, pp 1–58.
89. Ticku MK, Burch T: Purine inhibition of [3H]-gamma-aminobutyric acid receptor binding to rat brain membranes. Biochem Pharmacol 29:1217–1220, 1980.
90. Toffano G, Guidotti A, Costa E: Purification of an endogenous protein inhibitor of the high affinity binding of gamma-aminobutyric acid to synaptic membranes of rat brain. Proc. Nat Acad Sci 75:4024–4028, 1978.
91. Waddington JL, Cross AJ: Baclofen and muscimol; behavioral and neurochemical sequelae of unilateral intranigral administration and effects of 3H-GABA receptor binding. Naunyn-Schmiedeberg's Arch Pharmacol 306:275–280, 1979.
92. Watkins JC: Pharmacological receptors and general permeability phenomena of cell membranes. J Theoret Biol 9:37–50, 1965.
93. Willow M, Johnston, GAR: Barbiturates and calcium-activated adenosine triphosphatase. Neurosci Lett 14:361–364, 1979.
94. Yoneda Y, Kuriyama K: Presence of a low molecular weight endogenous inhibitor on 3H-muscimol binding in synaptic membranes. Nature (London) 285:670–673, 1980.

The Role of Peptides and Amino Acids as Neurotransmitters, pages 19–27
©1981 Alan R. Liss, Inc., 150 Fifth Avenue, New York, NY 10011

Excitatory Amino Acid Receptors

Hugh McLennan

Twenty years ago Curtis et al [1960] described the excitation and membrane depolarization of single mammalian central neurones which followed the extracellular administration of the naturally occurring acidic amino acids L-glutamate and L-aspartate, although the convulsant properties of glutamate had been recognized by Hayashi [1956] some years earlier. Shortly thereafter, Curtis and Watkins [1960] suggested that these and related compounds, which also caused excitation, reacted with a three-point receptor (Fig. 1) to induce the appropriate permeability changes. At that time, it was regarded as unlikely that the substances had physiological roles to play as excitatory synaptic transmitters: "When these amino acid ions were applied to the one interneurone, the intervals preceding the cessation of cell responses after the termination of the iontophoretic currents were virtually identical in each case, as indeed were the intervals found for both dextro- and laevo-forms of glutamate and aspartate ions. This suggests that enzymic removal of the applied acids was not a significant factor in determining the duration of these intervals, and since it is reasonable to consider that specific enzymes occur near excitatory synapses for the destruction of synaptically released transmitters, it follows that probably none of these amino acids plays a part as excitatory transmitter substance in the mammalian spinal cord" [Curtis et al, 1960]. A further objection was the apparent universality of action of the acidic amino acids in the sense that all neurones tested appeared to respond, in marked contrast to the situation with such more "classical" transmitter candidates as acetylcholine. Other early results also seemed to support the view that glutamate and aspartate were unlikely to act as transmitters of synaptic excitation [Curtis, 1965]. This situation has, of course, changed with the recognition that the cessation of amino acid action, as well as of other

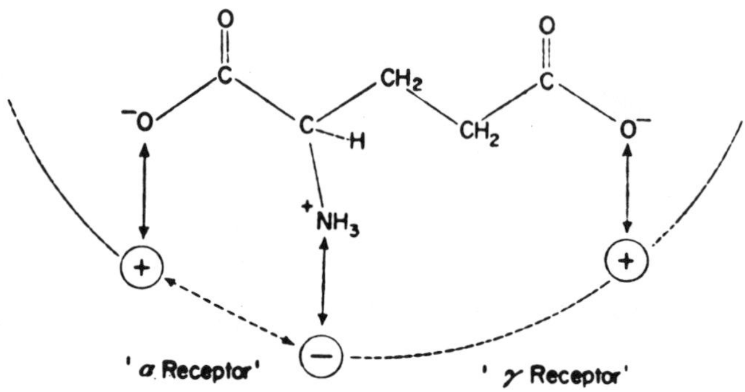

Fig. 1. The three-point attachment receptor proposed for glutamate. [Curtis and Watkins, 1960.]

recognized transmitter substances, occurs not through enzymic breakdown, but by processes of cellular uptake (see eg [Balcar and Johnston, 1972]), and the evidence for a prominent synaptic function for several amino acids both excitatory and inhibitory is now reasonably compelling. Indeed, glutamate and GABA between them may account for some 75% of all synaptic transmission processes occurring within the central nervous system of vertebrates (cf [Johnston], this volume).

There is, therefore, good reason to be concerned with the characteristics of the receptors with which the acidic amino acids react, both in connection with synaptic function and in that some compounds of the group have been reported to have toxicological properties. The original hypothesis of Curtis and Watkins [1960] of a three-point attachment site involving two cationic and one anionic groups for the receptors remains valid. However, there is now evidence that more than one such site exists, and this will be the principal subject for discussion in this paper.

Antagonists of the Excitatory Amino Acids

In the early 1970s, several compounds of diverse chemical structure were described as blockers of the excitatory effects of certain amino acids, although the degree to which their actions specifically affected glutamate was a matter of controversy. The substances implicated were L-glutamate diethylester (GDEE) (Fig. 2I) and α-methylglutamate [Haldeman et al, 1972],

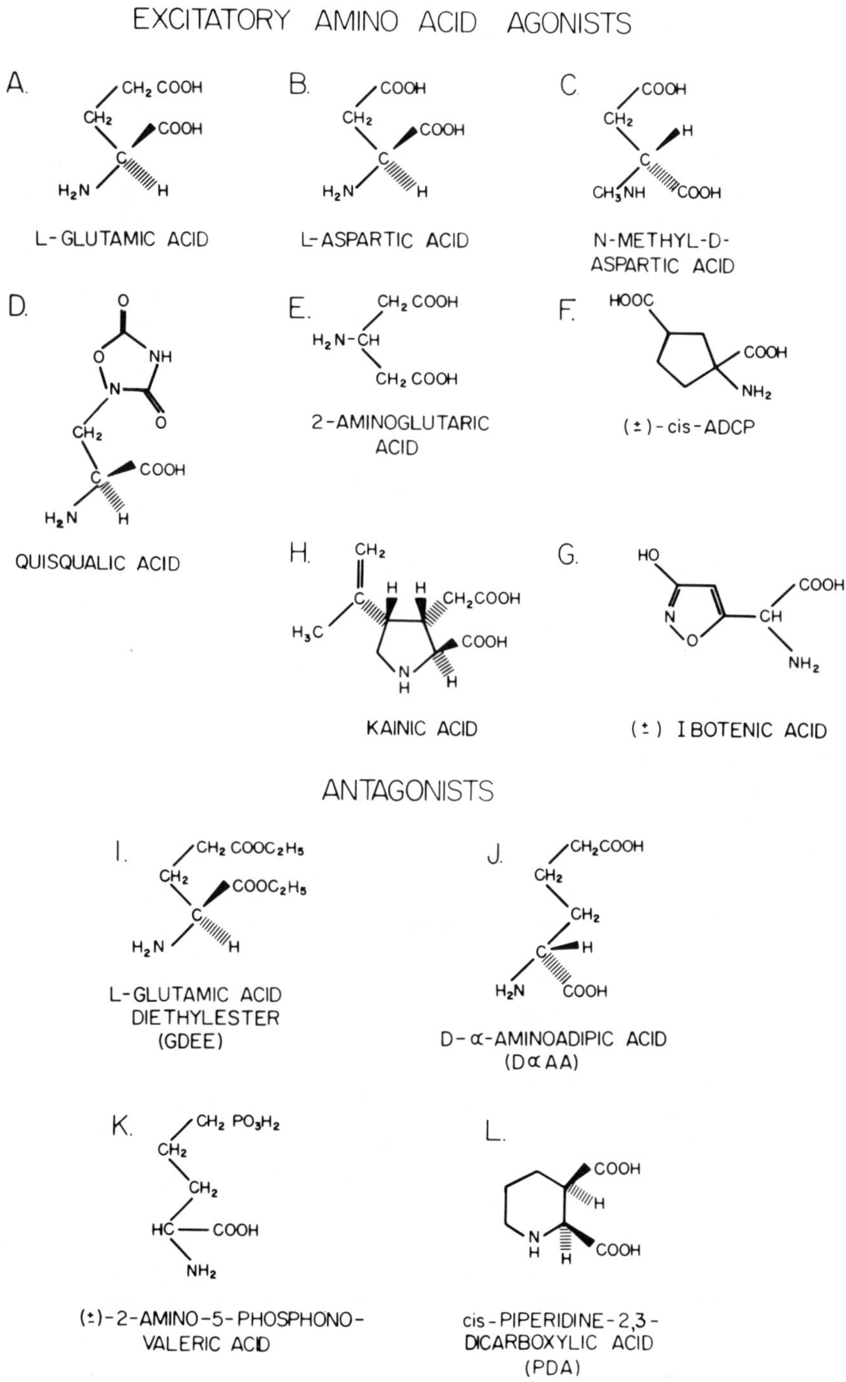
EXCITATORY AMINO ACID AGONISTS
A. L-GLUTAMIC ACID
B. L-ASPARTIC ACID
C. N-METHYL-D-ASPARTIC ACID
D. QUISQUALIC ACID
E. 2-AMINOGLUTARIC ACID
F. (±)-cis-ADCP
H. KAINIC ACID
G. (±) IBOTENIC ACID
ANTAGONISTS
I. L-GLUTAMIC ACID DIETHYLESTER (GDEE)
J. D-α-AMINOADIPIC ACID (DαAA)
K. (±)-2-AMINO-5-PHOSPHONO-VALERIC ACID
L. cis-PIPERIDINE-2,3-DICARBOXYLIC ACID (PDA)

Fig. 2.

D-lysergic acid diethylamide (LSD) [Boakes et al, 1970], L-methionine sulphoxamine and 9-methoxyaporphine [Curtis et al, 1972], l-hydroxy-3-aminopyrrolid-2-one [Davies and Watkins, 1973], and L-5,6-dimethoxy-aporphine [Duggan et al, 1973]. Subsequent work did not confirm the effectiveness of α-methylglutamate and LSD, but it was established that the other compounds depressed amino acid excitations, while having little or no effect on those induced by acetylcholine. The matter of their selectivity for one amino acid over another, however, was not so quickly settled, and Haldeman and McLennan [1972] claimed that GDEE could antagonize glutamate excitations selectively while Curtis et al [1972] failed to demonstrate such specificity of action.

The discrepancies appear to stem from several sources. In the first place, the potentially useful substances mentioned above do have "unspecific" actions if used in large enough quantities, ie, not only do amino acid-induced responses become attenuated, but the actions of acetylcholine and even the spontaneous activity of cells can be reduced. It is now recognized, however, that GDEE can have a selective antagonistic effect against glutamate excitations when this compound is compared with a number of others [Hicks et al, 1978; McLennan and Lodge, 1979], and a number of synaptic processes believed to be glutamate-mediated can also be antagonized (eg, [Haldeman and McLennan, 1972; Stone, 1973; Hicks and McLennan, 1979; Wheal and Miller, 1980]).

Continued work led, however, to the recognition that the reaction of glutamate and aspartate and their analogs with only a single class of receptor probably was incorrect. My colleagues and I were examining the relative potencies as excitants of the stereoisomeric pairs of some α-amino acids, and we confirmed the original work of Curtis and Watkins [1960], both in that D-glutamate was somewhat less active than the L form and in that the next higher homolog of L-glutamate, L-α-aminoadipate, was weaker still. D-α-Aminoadipate was not then available to us, but it was a considerable surprise to find that not only was the racemic mixture DL-α-aminoadipate not active as an excitant but that it also actually depressed the effects of other excitants including L-α-aminoadipate [Hall et al, 1977]. Subsequent work has confirmed the speculation which we made to the effect that D-α-aminoadipate (DαAA) (Fig. 2J) would prove to be an antagonist [Biscoe et al, 1977], but has also shown that its spectrum of action is quite different from that exhibited by GDEE [Hicks et al, 1978; McLennan and Lodge, 1979].

Table I sets forth the ranking of amino acids which occurs when they are listed in order of decreasing susceptibility to blockade by the two antagonists GDEE and DαAA. It will be observed that in the main, the two lists are converses one of the other, ie, those compounds whose excitatory actions are most affected by GDEE are least susceptible to DαAA and vice versa. It would appear, therefore, that at least two classes of receptors must exist, one preferentially blocked by DαAA and excited by the optical isomers of N-methylaspartate (NMLA and NMDA, Fig. 2C), the other antagonized by GDEE and characteristically excited by L-glutamate itself and by quisqualate (Fig. 2D). The possibility of a third type of receptor blocked by both antagonists seems less likely than the possibility that those agonists which appear in the middle of both lists (such as L-aspartate) are capable of some reaction with both receptor types.

Possible Configurations of the Receptors

Both ibotenate (Fig. 2G [Johnston et al, 1974]) and cis-l-amino-1,3-dicarboxycyclopentane (cis-ADCP, Fig. 2F [McLennan and Wheal, 1978]) were originally proposed as analogs of glutamate, yet, as judged by the data of Table I, both react preferentially with the DαAA-antagonized receptor. Glutamate itself possesses an extremely flexible molecule, while ibotenate and ADCP are rigid by reason of the planar ring structures to which the active substituents are attached, and thus the distances between those groups are relatively fixed in space and correspond to an extended form of glutamate. It was somewhat surprising in this case to find that these molecules appeared not to react with the same receptor with which glutamate preferentially combined, and this in turn would indicate that glutamate reacts with the GDEE-sensitive receptor in a relatively folded conformation. Those other molecules which also are preferentially blocked by GDEE (L-α-aminoadipate and the even more complex quisqualate) can assume folded conformations as well.

The spacing between the carboxyl and amino groups which are attached to the α-carbon atom of these amino acids is identical for all of them, and,

TABLE I. Relative Order of Antagonism of Amino Acid Excitation of Mammalian Neurones*

D-α-Aminoadipate	L-Glutamate diethylester
N-Methyl-D-aspartate	L-Glutamate
(±)-Ibotenate	Quisqualate
(±)-cis-ADCP	L-Aspartate
D-Glutamate	D-Glutamate
L-Aspartate	(±)-Ibotenate
Quisqualate	(±)-cis-ADCP
L-Glutamate	N-Methyl-D-aspartate

*Data from [Hicks et al, 1978; McLennan and Lodge, 1979].

therefore, the distance between the anionic and cationic attachment sites in the "α-receptor" of Figure 1 must be of the order of 0.25 nm. On the other hand, the distance between the distal carboxyl group and either the amino group or the α-carboxyl can vary rather widely. The spacing of the two groups reacting with the "γ-receptor" (Fig. 1) for rigid DαAA-antagonized molecules such as cis-ADCP would appear to be of the order of 0.5 nm, and molecules such as NMDA can almost achieve such a spacing in a fully extended form. At the folded GDEE-sensitive receptor, this distance in the γ-receptor will of course be less; but it is not clear why flexible molecules such as glutamate should prefer to react with a receptor in this conformation.

That it is the configuration of the γ-receptor and not the separation between the two anionic binding sites in the overall receptive template which determines whether it will be blocked by GDEE or by DαAA is revealed by examining the pharmacological sensitivity of the excitations produced by 2-aminoglutarate (Fig. 2E). Although this compound is the β-amino analog of glutamate, one of the carboxyl–amino spacings can fit the α-receptor portion of the template with an 0.25-nm spacing. In this case, the second carboxyl–amino moiety resembles aspartate while the carbon chain separating the two carboxyl groups is identical to that in the glutamate molecule. Therefore, if the intercarboxyl distance determines the receptor with which a molecule reacts, 2-aminoglutarate should be GDEE-sensitive as is glutamate, whereas if the γ-receptor portion is critical, 2-aminoglutarate will resemble aspartate by being relatively more sensitive to blockade by DαAA. This latter is experimentally found, and one may, therefore, conclude that the spacing between the cation-accepting and the distal anion-accepting sites determines the differential sensitivity of the receptors to the antagonists and constrains the molecular configuration of the reacting agonist molecules.

Other compounds related to DαAA have been proposed as amino acid antagonists by Watkins and his colleagues (see [Davies and Watkins, 1979] for references). These include 2-amino-5-phosphonovalerate, the phosphonic acid analog of α-aminoadipate (Fig. 2K); D-α-aminosuberate, a higher homolog with six methylene groups in the chain in contrast to the four of DαAA; and such diamino compounds as α, ϵ-diaminopimelate. All of these appear qualitatively to resemble DαAA in their pattern of action, although some, notably the aminophosphonovalerate compound, are considerably more potent. I have found, however, that one substance also proposed as an antagonist by Watkins has somewhat different properties, although like the others it does antagonize the actions of NMDA preferentially. This compound is cis-2, 3-piperidinedicarboxylate (PDA, Fig. 2L). Before this matter is further pursued, however, some consideration to the likely existence of yet another class of amino acid receptors must be presented.

Kainic Acid

The excitatory properties of this pyrollidine derivative (Fig. 2H) were described by Shinozaki and Konishi [1970], and its probable reaction with glutamate-preferring receptors was proposed by Johnston et al [1974]. In addition, it is powerfully neurotoxic ([Nadler, 1979] for a recent review). However, in studies in which the selectivity of action of GDEE and DαAA were compared, the behavior of kainate was aberrant in that the excitations induced by it were resistant to both antagonists, ie, its position would be near the bottom of both of the lists in Table I. This in turn implies that the reaction of kainate with either the GDEE- or with the DαAA-antagonized receptor types is not vigorous and that in all probability a third class of receptor must be involved.

Binding studies too indicate a distinction between the sites accepting kainate from those with which glutamate reacts. Thus, although glutamate can inhibit kainate binding to rat brain membranes [Johnston et al, 1979], the converse is not true, which indicates that kainate is unable to react with the glutamate-binding sites. The evidence, therefore, is for the existence also of a kainate-preferring receptor, but as kainate is not a constituent of the brain, the endogenous ligand for this receptor is presently unknown. However, the membrane fragments as normally prepared for binding studies contain an inhibitor of kainate binding analogous to the inhibitors of GABA binding (GABARINs) discussed by Professor Johnston elsewhere in this volume. That inhibitor of kainate binding could be the natural ligand.

Of the amino acid antagonists tested, only one, PDA, has been found to have a significant effect on kainate-induced excitations, a blocking action in fact comparable to its effects on NMDA. Its effects against glutamate and other amino acid agonists resemble those of DαAA. Thus, it would appear that PDA is probably capable of reaction with two of the receptor types, ie, with both the NMDA- and with the kainate-preferring classes.

Physiological Roles for the Excitatory Amino Acids

The ubiquitous distribution of glutamate and its near universality of action as an excitant render it probable that it together with GABA are quantitatively the most important transmitters within the vertebrate central nervous system, as was mentioned earlier. It is likely that glutamate, while capable of some reaction with the extended DαAA-sensitive receptor, prefers to activate the more folded GDEE-affected type, and there are recognized synaptic processes which are blocked by this glutamate antagonist, as has also been commented upon above.

The synaptic significance of the DαAA-blocked receptors is indicated by the reduction of responses in the spinal cord [Davies and Watkins, 1979] and hippocampus [Hicks and McLennan, 1979] produced by this compound, and at these sites aspartate is the more likely transmitter candidate given its greater sensitivity to antagonism by DαAA (Table I). The physiological role of the kainate receptors must remain uncertain as long as their natural ligand continues to be unidentified, although some are certainly extrasynaptic [Davies et al, 1979]. The relative importance of this class of receptor will also remain in doubt until a specific antagonist of kainate is found which can be tested on synaptic responses. That this is a feasible goal is indicated by the fact that PDA alone of the amino acid antagonists does have a significant effect against kainate.

Of the nature of the membrane changes produced by the various amino acid agonists and of the ionic permeabilities altered thereby, one has relatively little knowledge. On the basis of the changes in conductance induced, Lambert et al [1981] have distinguished three categories of amino acid agonists which agree well with the groups distinguished on the pharmacological grounds that I have discussed: Group 1, which shows fast on and off changes in membrane polarization accompanied by a moderate conductance increase, includes L-glutamate and quisqualate; Group 2, which has an effect upon membrane potential which is slower and associated with a conductance decrease, includes NMDA and ibotenate; and Group 3, which has kainate as its only member, is characterized by a slow depolarization associated with a large conductance increase. The changes induced in various membrane ionophores corresponding to these alterations in conductance are matters largely of speculation and of important future investigation.

SUMMARY

Professor Johnston has alluded earlier in this volume to the complexity of the GABA-activated systems. Although considerably less well understood, it is probable that the glutamate–aspartate receptors will prove to be as complicated and as diverse as those activated by GABA, and as quantitatively important. Given these indications, their involvement in the etiology of Parkinson's disease and other neurological disorders is eminently probable.

ACKNOWLEDGMENTS

The author's own work has been supported by the Medical Research Council of Canada. Gifts of compounds from and helpful discussions with Professor G.A.R. Johnston and Dr. J.C. Watkins are gratefully acknowledged.

REFERENCES

1. Balcar VJ, Johnston GAR: The structural specificity of the high affinity uptake of L-glutamate and L-aspartate by rat brain slices. J Neurochem 19:2657–2666, 1972.
2. Biscoe TJ, Evans RH, Francis AA, Martin MR, Watkins JC, Davies J, Dray A: D-α-Aminoadipate as a selective antagonist of amino acid-induced and synaptic excitation of mammalian spinal neurones. Nature (London) 270:743–745, 1977.
3. Boakes RJ, Bradley PB, Briggs I, Dray A: Antagonism of 5-hydroxytryptamine by LSD-25

in the central nervous system: A possible neuronal basis for the actions of LSD-25. Br J Pharmacol 40:202–218, 1970.

4. Curtis DR: The actions of amino acids upon mammalian neurones. In Curtis DR, McIntyre AK (eds): "Studies in Physiology." New York: Springer, 1965, pp 34–42.
5. Curtis DR, Duggan AW, Felix D, Johnston GAR, Tebecis AK, Watkins JC: Excitation of mammalian neurones by acidic amino acids. Brain Res 41:283–301, 1972.
6. Curtis DR, Phillis JW, Watkins JC: The chemical excitation of spinal neurones by certain acidic amino acids. J Physiol (London) 150:656–682, 1960.
7. Curtis DR, Watkins JC: The excitation and depression of spinal neurones by structurally related amino acids. J Neurochem 6:117–141, 1960.
8. Davies J, Evans RH, Francis AA, Watkins JC: Excitatory amino acid receptors and synaptic excitation in the mammalian central nervous system. J Physiol (Paris) 75:641–644, 1979.
9. Davies J, Watkins JC: Microelectrophoretic studies on the depressant action of HA-966 on chemically and synaptically excited neurones in the cat cerebral cortex and cuneate nucleus. Brain Res 59:311–322, 1973.
10. Davies J, Watkins JC: Selective antagonism of amino acid-induced and synaptic excitation in the cat spinal cord. J Physiol (London) 297:621–635, 1979.
11. Duggan AW, Lodge D, Biscoe TJ, Headley PM: Effect of nuciferine on the chemical excitation of Renshaw cells in the rat. Archs Int Pharmacodyn Ther 204:147–149, 1973.
12. Haldeman S, Huffman RD, Marshall KC, McLennan H: The antagonism of the glutamate-induced and synaptic excitations of thalamic neurones. Brain Res 39:419–425, 1972.
13. Haldeman S, McLennan H: The antagonistic action of glutamic acid diethylester towards amino acid-induced and synaptic excitations of central neurones. Brain Res 45:393–400, 1972.
14. Hall JG, McLennan H, Wheal HV: The actions of certain amino acids as neuronal excitants. J Physiol (London) 272:52–53P, 1977.
15. Hayashi T: "Chemical Physiology of Excitation in Muscle and Nerve." Tokyo: Nakayama–Shoten, 1956.
16. Hicks TP, Hall JG, McLennan H: Ranking of excitatory amino acids by the antagonists glutamic acid diethylester and D-α-aminoadipic acid. Canad J Physiol Pharmacol 56:901–907, 1978.
17. Hicks TP, McLennan H: Amino acids and the synaptic pharmacology of granule cells in the dentate gyrus of the rat. Canad J Physiol Pharmacol 57:973–978, 1979.
18. Johnston GAR, Curtis DR, Davies J, McCulloch RM: Spinal interneurone excitation by conformationally restricted analogues of L-glutamic acid. Nature (London) 248:804–805, 1974.
19. Johnston GAR, Kennedy SME, Twitchin B: Action of the neurotoxin kainic acid on high affinity uptake of L-glutamic acid in rat brain slices. J Neurochem 32:121–127, 1979.
20. Lambert JDC, Flatman JA, Engberg I: Actions of excitatory amino acids on membrane conductance and potential in motoneurones. In Di Chiara G, Gessa GL (eds): "Glutamate as a Neurotransmitter." New York: Raven Press, 1981, pp 205–216.
21. McLennan H, Lodge D: The antagonism of amino acid-induced excitation of spinal neurones in the cat. Brain Res 169:83–90, 1979.
22. McLennan H, Wheal HV: A synthetic, conformationally restricted analogue of L-glutamic acid which acts as a powerful neuronal excitant. Neurosci Letts 8:51–54, 1978.
23. Nadler JV: Kainic acid: Neurophysiological and neurotoxic actions. Life Sci 24:289–300, 1979.
24. Shinozaki H, Konishi S: Actions of several anthelminthics and insecticides on rat cortical neurones. Brain Res 24:368–371, 1970.
25. Stone TW: Cortical pyramidal tract interneurones and their sensitivity to L-glutamic acid. J Physiol (London) 233:211–225, 1973.
26. Wheal HV, Miller JJ: Pharmacological indentification of acetylcholine and glutamate excitatory systems in the dentate gyrus of the rat. Brain Res 182:145–155, 1980.

The Role of Peptides and Amino Acids as Neurotransmitters, pages 29–52
©1981 Alan R. Liss, Inc., 150 Fifth Avenue, New York, NY 10011

Interactions of Opioid Peptides With Their Receptors

Jacob M. Hiller

The relatively slow progress in elucidating the molecular mechanisms underlying the myriad of pharmacological effects of morphine and its congeners may lie in part in our, as yet, incomplete understanding of the operations of the target issue, the central nervous system. However, developments in the neurochemical and neuropharmacologic aspects of opiate research in the past 8 years have engendered considerable activity and excitement. Two major breakthroughs, the discovery of opiate receptors in the central nervous system of animals and man and the ensuing discovery of endogenous opiate-like ligands for these receptors, are the focal points of this chapter.

Pharmacological evidence supporting the hypothesis that opiates must bind to highly specific sites on nerve cells in order to elicit their responses has been available for several decades. The process of binding is presumed to initiate a cascade of chemical or physical events leading to the observed responses. The binding site together with its transducing mechanism was termed a receptor in line with the nomenclature used in the field of hormone study.

The high degree of structural and steric specificity that characterizes the action of opiates lent credence to this receptor postulate. Thus, for a large number of morphine-like analgesics studied, it is only the levorotatory enantiomer that is active, while the dextrorotatory isomer has little or no analgesic or addiction-producing activity. Small modifications of chemical structure of these compounds can result in profound changes in pharmacological potency. The most studied structural change is the substitution of the tertiary nitrogen-methyl group by a larger alkyl group. Such a substitution frequently results in the formation of a molecule with potent specific antagonist activity against many of the actions of morphine and related narcotic analgesics. So, the N-allyl and N-cyclopropylmethyl analogs of oxymorphone are the so-called "pure" antagonists, naloxone and naltrexone. The easiest explanation for antagonist activity is competition with an agonist for a receptor site. The binding of an antagonist is assumed not to produce the subsequent changes needed to evoke a response. However, N-methyl substitution may also lead to the formation of compounds which have mixed agonist–antagonist properties, such as nalorphine which is the N-allyl analog of morphine. This finding has led to the synthesis of many compounds with mixed pharmacological profiles in the hope of finding a potent opiate analgesic devoid of addiction liability. These requirements of steric and structural specificities led to the working hypothesis that opiates must bind to highly specific macromolecular sites or receptors in order to initiate their pharmacological effects.

Discovery of Opiate Receptors

Biochemical identification of opiate receptors proved difficult because of the problem that initially faces any receptor binding study, namely, distinguishing specific binding to receptors from nonspecific binding to various tissue components. Ingoglia and Dole [1970] were the first investigators to use the principal of stereospecificity in an attempt to identify opiate receptors. They sought, though unsuccessfully, to find a difference in the rate of diffusion into rat brain of the enantiomers, d- and l-methadone, after their injection into the lateral ventricles. In 1971, Goldstein et al used this criterion of stereospecificity in their attempts to identify opiate binding sites in mouse brain homogenate. They incubated the homogenate with [^{3}H]levorphanol in the presence of a large excess of unlabeled dextrorphan or levorphanol. Stereospecific binding was defined as that portion of the binding that was prevented by levorphanol but not by dextrorphan. However, in their experiments, only on the average of 2% of the total binding appeared to be stereospecific.

Incontrovertable evidence for the existence in rat brain of stereospecific binding sites for opiates, which represented the major portion of total binding, was independently and simultaneously reported by three laboratories [Simon et al, 1973; Terenius, 1973a; and Pert and Snyder, 1973]. These investigative groups all modified the procedure of Goldstein et al [1971] in very similar ways so as to minimize nonspecific binding. These modifications involved the use of tritiated opiate ligands of high specific activity which made it possible to use very low concentrations of radioactive ligand, and the rapid washing with cold buffer of filters after the bound homogenate was filtered, or of pellets of bound homogenate following centrifugation. This second modification reduced dramatically the amount of unbound or loosely bound radioactivity remaining on filters or pellets. Results from experiments using these techniques demonstrated that 50 to 90% of the observed opiate binding was stereospecific in nature. This radioreceptor assay has since that time become a standard procedure in laboratories around the world for the in vitro demonstration of opiate receptors. These receptors are ubiquitous in the higher animals, having been found in man [Hiller et al, 1973], in all vertebrates so far studied [Pert et al, 1974a], and, most recently, in some invertebrates [Pert and Taylor, 1979; Stephano et al, 1980].

Characterization of Opiate Receptors

Stereospecific opiate binding sites are found in the central nervous system and in the innervation of smooth muscle systems, such as the guinea pig ileum and the mouse vas deferens. They are tightly bound to membraneous elements of nerve cells and have been reported to be most abundant in the synaptosomal fraction of brain and guinea pig ileum homogenates [Terenius, 1973b; Pert et al, 1974b; Hitzeman et al, 1974]. The stereospecific binding is saturable, and at saturation amounts to about 0.35 pmol of bound ligand per milligram of protein in a crude membrane preparation. Dissociation constants range from 0.025 nM for certain derivatives of fentanyl [Stahl et al, 1977] to a low of 10 μM for codeine [Creese and Snyder, 1975]. The fact that levorphanol and dextrorphan differ in their affinities by four orders of magnitude [Simon et al, 1974; Pert and Snyder, 1973] is the best example of the remarkable ability of the opiate receptor to distinguish between stereoisomers.

The pH optimum for opiate binding, 6.5–8, lies within the physiological range. Binding is decreased by alkali, alkaline earth, transition metal, and organic cations. An important exception is the sodium ion and to a lesser extent the lithium ion, which at concentrations of up to 150 mM enhance antagonist and depress agonist binding. This discriminatory effect of sodium will be discussed later in this review.

The suggestion that protein is involved in the opiate binding site is supported by the finding that stereospecific binding is inhibited by proteolytic agents [Simon et al, 1973; Pasternak and Snyder, 1973] and by a variety of protein reagents [Simon et al, 1973; Terenius, 1973b]. The participation of phospholipids in the receptor is not as clear, since binding is blocked by phospholipase A from *Vipera russelli* or bee venom [Pasternak and Snyder, 1973], but not by phospholipase A from *Crotalus adamanteus* or by phospholipase C [Simon et al, 1973]. Ribonuclease, deoxyribonuclease, and neuraminidase are ineffectual [Pasternak and Snyder, 1973].

One of the most important and convincing pieces of evidence that we are truly dealing with the pharmacologically relevant opiate receptor is the excellent correlation seen between in vitro affinities of a large number of opiate ligands and the effective dosages of these same drugs in eliciting analgesia in whole animals [Wilson et al, 1975; Stahl et al, 1977] or in inhibiting the electrically induced contractions of the guinea pig ileum [Creese and Snyder, 1975].

The stereospecificity, saturability, high affinity of opiate binding, and the excellent correlation between pharmacological potency and binding affinity of opiates and their antagonists, strongly support the hypothesis that these sites are the pharmacologically relevant recognition and binding sites of opiate receptors.

Regional Distribution of Opiate Receptors in the CNS

The precise pattern of distribution of stereospecific opiate binding sites in the CNS provides information as to the anatomical sites of opiate action. Our laboratory carried out a detailed study [Hiller et al, 1973] of the regional distribution of [^{3}H]etorphine binding in human brain obtained at autopsy from the office of the Chief Medical Examiner of the City of New York. Studies in monkey brain yielded very similar results [Kuhar et al, 1973]. As would be expected, the receptors are not uniformly distributed throughout the CNS. Many areas of the CNS, which in this distribution study demonstrated high levels of binding, were shown to be part of, or associated with, the limbic system and/or implicated in pain conduction and modulation, eg, periventricular and periaqueductal gray areas, thalamus, substantia gelatinosa of the dorsal cord, nucleus raphe magnus, and nucleus reticularis paragigantocellularis. The limbic system has long been thought to be the target tissue for morphine's action in ameliorating the affective aspects of the perception of pain. These results have been confirmed and extended in a series of autoradiographic studies of opiate receptor distribution in rat CNS [Atweh and Kuhar, 1977a,b,c].

Conformational Changes in Opiate Receptors

Earlier in the discussion, it was mentioned that in the presence of sodium ions, agonist binding is inhibited and antagonist binding is augmented. This finding arose from an experimental discrepancy between two laboratories. The Simon group reported [Simon et al, 1973] that sodium inhibits the binding of [^{3}H]etorphine, while Pert and Snyder [1973] reported the binding of [^{3}H]naloxone to be insensitive to sodium chloride. We then suggested [Simon et al, 1973] that the apparent discrepancy might represent a general difference in the manner in which agonists and antagonists bind to the receptor. This remarkable discrimination by a small ion between closely related molecules (eg, morphine and nalorphine, oxymorphone and naloxone) is highly specific in that only sodium, and to a lesser extent lithium, affects differentially the binding of opiates. Studies on the mechanism of the sodium effect were carried out and the conclusions were drawn that sodium increased the affinity of the receptor for antagonists while reducing the affinity for agonists [Simon et al, 1975a]. Pert and Snyder [1974] held that sodium increased the number of high affinity binding sites for antagonists and decreased the number of binding sites for agonists. Both laboratories also reported [Pert and Snyder, 1974; Simon et al, 1975a] the results of experiments in which unlabeled opiates were allowed to compete with labeled antagonists in the presence and absence of sodium. When the unlabeled competitor was a relatively pure antagonist, there was little or no change in the ED_{50} (concentration able to displace the binding of 50% of the labeled drug), since both competitor and labeled antagonist have similar shifts in affinity in the presence of sodium. However, the ED_{50} of agonists, in the presence of sodium, increased enormously (10- to 60-fold) and the increase with mixed agonist–antagonist opiates was intermediate. A shift in ED_{50} reflects a change in affinity and not a change in the number of binding sites.

Despite the differences between laboratories in their experimental results, both groups have independently proposed similar models with which to view the mechanism of the sodium effect [Simon, 1974; Pert and Snyder, 1974; Simon, 1975]. The models propose that the sodium ion acts as an allosteric effector, the binding of which to an allosteric site causes a conformational change of the opiate binding site. The new sodium-dependent conformer exhibits a higher affinity for antagonists and a lower affinity for agonists.

A study on the kinetics of receptor inactivation by sulfhydryl reagents [Simon and Groth, 1975] led to independent evidence for the conformational change in opiate receptors produced by sodium. As stated earlier, opiate receptor binding is inactivated by reagents which alkylate SH-groups, such as N-ethylmaleimide (NEM). The temperature-dependent rate of inactivation by NEM followed pseudo-first-order kinetics and was equally effective against antagonist and agonist binding. However, when crude brain membrane homogenates were incubated with NEM in the presence of 100 mM NaCl, the half-time of inactivation increased from 10–30 min. An effect of NaCl on the alkylation was ruled out, since it was not observed when model SH-compounds, such as glutathione, were reacted with NEM. In addition to sodium, only lithium exhibited, but to a lesser degree, any protective effect. It was, therefore, postulated that the protective effect of sodium against the inactivation of receptors via alkylation of SH-groups is the result of a conformational change which renders the SH-group less accessible to NEM and is identical to the conformational change which results in enhanced antagonist and reduced agonist binding.

Progress in Receptor Solubilization

The solubilization of an active opiate receptor will permit us to determine the chemical composition and subunit structure of the receptor molecule. In addition, it will enable us to produce antibodies to the receptor which upon injection into animal brain may clarify the physiological role of opiate receptors and their ligands.

The tight association of opiate receptors with cell membranes requires a method of solubilization that will permit the receptors to remain in an active state. This has proven to be a difficult problem, since a major impediment has been the inactivation of opiate receptors by extremely low concentrations of detergents commonly used to solubilize membrane proteins. Several years ago, our laboratory was successful in solubilizing an etorphine-receptor complex with the nonionic detergent Brij 36T [Simon et al, 1975b]. However, even with this detergent, dissociation of etorphine was more rapid from the solubilized complex than from membrane-bound receptors, and stereospecific binding of opiates to the solubilized macromolecule, after etorphine dissociation, was not preserved.

We have recently been successful in solubilizing active opiate receptors from the brain of toad *(Bufo marinus)* using 1% digitonin [Ruegg et al, 1980]. The solubilization of these receptors using digitonin was based on previous work with β-adrenergic receptors. Success in the solubilization of β-adrenergic receptors in active form using digitonin was attained from nonmammalian sources [Caron and Lefkowitz, 1976; Vauquelin et al, 1977] but not from mammalian sources [Strauss et al, 1979].

In toad brain, saturable stereospecific binding of [^{3}H]diprenorphine, representing 50–60% of total binding, was obtained in the soluble fraction. Up to 30–40% of the original stereospecific binding activity and 20–30% of the membrane protein had been solubilized from the membranes. As had been found in the membrane-bound receptor, trypsin, NEM, and heat readily inactivated binding activity of the solubilized receptor, indicating the protein nature of the receptor. Using gel filtration, the solubilized receptor has been purified 6–8-fold.

Recently two other laboratories published accounts of the solubilization of active opiate receptors. Using a new type of detergent, which is a zwitterionic derivative of cholic acid, Simonds et al [1980] have solubilized active opiate receptors from neuroblastoma × glioma hybrid cell NG 108-15 membranes and have also reported some success with rat brain membranes. Bidlack and Abood [1980] have reported the solubilization of opiate binding sites from rat brain using Triton X-100. Attempts, in our and other laboratories, to solubilize rat brain opiate receptor with this nonionic detergent have been unsuccessful.

Organotypic Culture System for the Study of Opiate Mechanisms

Model systems for the bioassay of opiates and, as we shall later discuss, opioid peptides, utilizing in vitro preparations of guinea pig ileum [Paton, 1957] and mouse vas deferens [Hughes et al, 1975a] have been developed. Proof as to the value of these models is seen in the close correlation between the efficacy of opiate agonists and opioid peptides to depress electrically induced contractions in these systems and the pharmacological potencies of these substances in man [Kosterlitz and Waterfield, 1975]. This depressive effect on the bioelectric activities of these preparations can be reversed or blocked by opiate antagonists, thus implying a mechanism involving opiate receptors.

In addition to these model preparations from the peripheral nervous system, an in vitro system consisting of an organotypic culture of specific central nervous system tissue is being utilized to study opiate mechanisms [Crain et al, 1977, 1978, 1979]. This preparation consists of spinal cord explants with attached dorsal root ganglia (DRG) derived from 14-day-old mouse fetuses [Crain and

Peterson, 1974, 1975]. The explants are grown in Maximow depression slide chambers in a serum supplemented nutrient culture medium containing nerve growth factor.

Focal stimulation of the DRG elicits complex negative slowwave potentials restricted to dorsal regions of the spinal cord explant and resembles primary and secondary sensory evoked synaptic network responses observed in situ in dorsal spinal cord [Crain and Peterson, 1974]. Marked and sustained depression of the major components of the DRG-evoked negative slow-wave responses in the dorsal cord are observed upon the introduction into the fluid bathing the cultures of analgesic concentrations of: morphine (10^{-7}–10^{-6} M), levorphanol (10^{-7} M), or etorphine (10^{-8}–10^{-7} M). Dextrorphan was without effect at a concentration of 10^{-6} M [Crain et al, 1977]. The potencies of opiate agonists in this system correlate well with the analgesic efficacy of these compounds in intact animals. Restoration of opiate-blocked cord responses occurred within minutes after the addition of low concentrations of naloxone (10^{-8}–10^{-6} M).

The electrophysiological effectiveness of opiates led us to study opiate receptor levels and their distribution in this model system [Hiller et al, 1978a,b]. Determination of opiate receptor binding levels were made in spinal cord-DRG cultures as well as in cultures of deafferented cord explants and isolated DRGs. Receptor binding was assayed using the potent opiate antagonist, [^{3}H]diprenorphine. Due to the stimulatory effect of nerve growth factor in the culture media, a profuse neuritic outgrowth of DRG neurons develops and extends for several millimeters beyond the explant area. At times, separate binding assays were carried out on explant and outgrowth zones. Stereospecific opiate binding is found in homogenates of these cultures. Freshly explanted material exhibits little or no binding; however, receptor levels increase with time and reach maximum levels after 11–14 days in vitro. The neuritic outgrowth of both isolated DRG cultures and cord-ganglion cultures contains the greatest amount of opiate binding (40–300 fmole/mg of protein).

The central branches of DRG neurons have been shown to establish synaptic connections within the dorsal horn of cocultured spinal cord explants [Crain and Peterson, 1975]. The presence of high levels of opiate receptors in the neuritic outgrowth from DRG explants grown in isolation would, therefore, suggest a location on the presynaptic primary afferent fibers in the spinal cord. This, in fact, is the first direct evidence for a presynaptic location of opiate receptors. Presynaptic location of opiate receptors was initially suggested by the finding that the receptor level in the dorsal horn of adult monkey spinal cord was reduced to about 50% of control values after dorsal rhizotomy [Lamotte et al, 1976]. These authors attributed this decrease to degeneration of opiate receptor bearing afferent terminals. Further binding studies in our cord-DRG explants [Hiller et al, 1978a] gave results which confirmed this finding. Dorsal cord from DRG innervated cultures had binding levels 150% above the level seen in the dorsal portion of cords grown in isolation (deafferented cords). No changes are seen in the levels of binding in the ventral portion of these cords. These data, taken together with the results from whole animal studies, strongly argue for the presence of opiate receptors on the primary afferents that originate in DRG neurons.

Discovery of Endogenous Opioid Peptides

When a survey of possible interactions between known neurotransmitters and neuromodulators with the opiate receptor proved negative, the possibility was considered that an, as yet, unknown opioid ligand was present in the CNS. This proposal received support from the finding that analgesia can be produced in the rat by electrical stimulation of the mesencephalic central gray and the periventricular gray matter [Liebeskind et al, 1974; Mayer and Liebeskind, 1974]. This analgesia was found to exhibit cross-tolerance with morphine-induced analgesia [Mayer and Hayes, 1975] and was partially reversed by the administration of naloxone [Akil et al, 1976]. These results are most readily explained by the existence of an endogenous opioid substance that can be released by electrical stimulation.

John Hughes [1975a,b] first reported the discovery of such an endogenous factor in the aqueous extracts of porcine brain. These extracts demonstrated opiate-like properties in bioassay systems. The isolation of a water-soluble material from rat and calf brain which competed for binding in opiate radioreceptor assays was accomplished by Terenius and Wahlstrom [1974; 1975]. Both materials, as well as an opioid substance extracted by Pasternak et al [1975] from calf brain, appeared very similar. These endogenous opioid substances were reported to be water soluble, heat stable, degradable by peptidases and to range in molecular weight between 800 and 1200 daltons. Goldstein and collaborators [Teschemacher et al, 1975; Cox et al, 1975] then described the isolation of a larger opioid factor from bovine pituitary as well as from crude preparations of ACTH.

Two pentapeptides, H-Tyr-Gly-Gly-Phe-Met-OH and H-Tyr-Gly-Gly-Phe-Leu-OH, were purified from the extracts of pig brain [Hughes et al, 1975b] and were proven to be active opioid substances. They were given the appellation methionine (Met) and leucine (Leu) enkephalin (from the Greek, meaning: in the head) by this group.

The observation was made [Hughes et al, 1975b] that the amino acid sequence of Met enkephalin was identical to residues 61–65 of the hormone β-lipotropin (β-LPH), a hormone first isolated from the pituitary gland by Li [1964]. This finding led rapidly to the uncovering of three longer peptides with opioid activity, all representing sequences present in β-LPH, ie, α-(β-LPH_{61-76}). β-(β-LPH_{61-91}) and γ-(β-LPH_{61-77}) endorphin [Li and Chung, 1976; Bradbury et al, 1976; Cox et al, 1976; Guillemin et al, 1976]. β-Endorphin, the 31 amino acid C-terminal fragment of β-LPH, proved to be the most important of these larger opioid peptides. Both α- and γ-endorphin may be breakdown products of β-endorphin or artifacts of its isolation [Rossier et al, 1977]. The now widely accepted generic term, endorphin, was suggested by E. J. Simon to signify all endogenous substances with morphine-like activity.

Recently, additional peptides with opioid activity have been reported. One of the most important of these is a peptide from the pituitary characterized by Goldstein et al [1979]. Because of its potent opioid activity in bioassay systems, these investigators named this peptide dynorphin. Dynorphin has been found in certain areas of the CNS in addition to the pituitary gland.

Heterogeneity of Opiate Receptors

With the introduction of the subject of endorphins, the topic of opiate receptor heterogeneity may now be discussed briefly. The concept of multiple receptors for neurotransmitters is not novel, since heterogeneity of receptors has long been established for cholinergic, adrenergic, and histamine receptors.

The noticeable differences in pharmacological responses to different types of narcotic analgesics including varying susceptibility to naloxone antagonism, differing pharmacological profiles, and their inability to substitute for each other in suppression of withdrawal symptoms in the addicted chronic spinal dog led Gilbert and Martin [1976] to postulate that three opiate receptor types are present in the CNS. These were named for the prototype drug that gave rise to the distinction: μ for morphine, κ for ketocyclazocine, and σ for SKF 10047.

Convincing data suggesting that the opiate receptors in mouse vas deferens are different from those present in the guinea pig ileum have been published [Lord et al, 1977]. This, together with evidence from binding competitions in guinea pig brain homogenates, was taken by this group as evidence that there exists a morphine-preferring receptor (μ) and an enkephalin-preferring receptor (δ). In these competitions, when labeled opiates were used, unlabeled opiates competed more effectively for binding than enkephalins, while the unlabeled enkephalins were more effective competitors than opiates against labeled enkephalins. This laboratory has confirmed these results [Simon et al, 1980] which indicate that at least two different types of opiate receptors exist, with overlapping specificity but with affinities that favor different ligands. Independent biochemical verification of μ- and δ-receptor entities have been presented by Robson and Kosterlitz [1979] and Smith and Simon [1980].

Experimental Approaches and Methods Used to Study Endorphins

In order to define the physiological role of the endorphins, a number of approaches have been used. These include (a) quantitative and qualitative mapping of the distribution and determining how this pattern is affected by physiological and environmental factors, (b) assessment of the pharmacological effects caused by the administration of endorphins to animal and human subjects, (c) ascertainment of the effects of narcotic antagonist administration to counter the effects of endorphins, and (d) investigations into the biosynthesis and metabolism of the endorphins. The results of these approaches will be summarized in the remaining text.

Methodology for the identification and quantitation of endorphins requires both high specificity and sensitivity. The earliest methods used, the radioreceptor assay and the bioassay systems can both detect the presence of opioid peptides. In the sensitive radioreceptor assay, the opioid activity present in a tissue extract can be measured by the extent to which it can compete for receptor binding with labeled opiates. In the less sensitive bioassays, in which opioid peptides can inhibit electrically induced contractions, the specificity of the inhibition can be confirmed by its reversibility by narcotic antagonists. Both methods suffer from the disadvantage of not distinguishing between different active endorphins. In addition, the presence of degradative enzymes in brain homogenates or nonendorphin inhibitors in tissue extracts might render uncertain the results of opiate receptor binding assays.

Assays that make use of highly specific antibodies against Met- and Leu-enkephalin and β-endorphin have proven to be most specific and sensitive. The radioimmunoassay (RIA), which measures the amount of peptide present in a tissue or plasma extract by competition with a radiolabeled peptide for binding to a given antibody, can detect femtomoles of peptide. The problem of immunological cross-reactivity can be overcome by separation of the peptides via high-pressure liquid chromatography prior to carrying out the RIA [Meek and Bohan, 1978; Rubinstein et al, 1977].

Finally, for mapping the distribution of opioid peptides, the immunohistochemical techniques have proven to be extremely useful. In this method, freeze-dried slices of tissue are exposed to antiserum to a particular peptide. Visualization of the antigen–antibody complex is accomplished by the use of a second antibody (against the first) conjugated with a fluorescent marker or by utilization of the peroxidase–antiperoxidase method.

Anatomical Distribution of Endorphins

Using RIAs and immunohistological techniques, detailed mappings of the distribution of enkephalins [Elde et al, 1976; Hökfelt et al, 1977; Simantov et al, 1977; Yang et al, 1977; Watson et al, 1977a] and β-endorphin [Rossier et al, 1977; Watson et al, 1977b; Bloom et al, 1978] have been carried out. Their results are summarized below.

Enkephalins have been found in all vertebrates so far studied [Simantov et al, 1976] and in some invertebrates [Alumets et al, 1979; Zipser, 1980]. They are widely distributed throughout the brain and spinal cord. No detectable differences in the distribution of Leu- and Met-enkephalin immunoreactivity have been found. Met-enkephalin levels are higher than those of Leu-enkephalin, though the ratio does vary between brain regions. In general, the distribution of enkephalins parallels that of the opiate receptor. However, notable exceptions do exist. The globus pallidus, which has the highest level of enkephalin immunoreactivity in rat brain, is relatively low in opiate receptors, and the caudate nucleus, which possesses a high concentration of opiate receptors, reveals only a moderate amount of enkephalin immunoreactivity.

Areas of high enkephalin concentration are found in the limbic system and include the hypothalamus and the periventricular, ventromedial, dorsomedial, and medial preoptic nuclei in the floor of the hypothalamus and the infundibulum. The regions involved in the transmission and modulation of pain, such as the central gray region, nucleus raphe magnus, nucleus reticularis giganto-cellularis, nucleus caudalis of the spinal tract of the trigeminal nerve, and the substantia gelatinosa of the spinal cord, are similarly areas of high enkephalin concentration. Neither rhyzotomy nor spinal transsection causes depletion of the enkephalin content of the subtantia gelatinosa, leading to the suggestion that enkephalinergic terminals may be located primarily on short interneurons rather than on the long ascending and descending pathways. In addition to finding the two pentapeptides in nerve terminals, techniques involving inhibition of axonal transport with colchicine have permitted the visualization of enkephalin containing cell bodies. It is estimated there may be 20 or more separate groups of enkephalin containing cells in the brain.

The distribution of β-endorphin is quite different from that of the enkephalins. β-Endorphin is restricted to a single group of cell bodies in the arcuate nucleus of the basal hypothalamus and is carried by processes emanating from this region to the midline areas near the ventricular surfaces. Thus β-endorphin containing nerve tracts are seen projecting from the basal hypothalamus forward through the preoptic area, around the anterior commisure, and into the periaqueductal region of the diencephalon and pons. Midline structures containing β-endorphin include the anterior paraventricular nucleus, the dorsal raphe, and the locus coeruleus. The distinctive distribution patterns of β-endorphin and the enkephalins in the CNS implies separate physiological roles for these peptides.

High concentrations of β-endorphin are found in the pars intermedia and pars distalis of the pituitary gland, with lower concentrations in the pars nervosa [Bloom et al, 1977]. Hypophysectomy fails to alter the levels of endorphins in the brain [Cheung and Goldstein, 1976], implying that endorphins found in the brain are not of pituitary origin.

Opioid peptides are also located outside of the CNS. In the gastrointestinal system, one of the targets of opiate action, enkephalin immunoreactive fibers are found throughout the tract with highest concentrations in the stomach, duodenum, and rectum. In the guinea pig ileum, enkephalin immunoreactive fibers mainly appear in the circular muscle layer and in Meissners plexus [Polak et al, 1977; Linnoila et al, 1978]. Enkephalins have also been extracted from sympathetic ganglia [Di Guilio et al, 1978] and enkephalin-like immunoreactivity has been seen in the adrenal medulla [Schultzberg et al, 1978] and in the neuroblastoma x glioma cell line 108CC15 [Glaser et al, 1980], a hybrid cell line containing opiate receptors. β-Endorphin immunoreactivity has been reported present in human pancreatic [Bruni et al, 1979] and placental tissue [Houck et al, 1980].

Pharmacology of the Endorphins

Intracerebral or intraventricular injection of opioid peptides elicits various characteristic effects found with the administration of opiate alkaloids. These include analgesia [Buscher et al, 1976; Belluzzi et al, 1976; Tseng et al, 1976; Bradbury et al, 1977], hypothermia [Holaday et al, 1978], respiratory depression [Florez and Mediavella, 1977], antidiuretic effects [Tseng et al, 1978], changes in electroencephalogram patterns [Pfefferbaum et al, 1978], and a variety of behavioral changes [Kastin et al, 1976, 1978; Katz and Roth, 1979] including the induction of catatonia [Bloom et al, 1976].

The hope that the endorphins would produce desired effects of morphine, such as analgesia, without the development of tolerance and dependence was quickly dashed. Physical dependence on Met-enkephalin and β-endorphin was apparent after their infusion into the periaqueductal gray-fourth ventricular spaces of rat brain [Wei and Loh, 1976]. The endorphins can substitute for morphine in the suppression of withdrawal symptoms in morphine-dependent animals [Mello and Mendelson, 1978]. Cross tolerance to morphine has been demonstrated [Waterfield et al, 1976].

Our organotypic cultures of fetal mouse spinal cord with attached dorsal root ganglia develop tolerance to opiate alkaloids and opioid peptides [Crain et al, 1979]. This is the first report that tolerance can develop to the classical opiates or to the endorphins in isolated explants of CNS tissue. Cross tolerance between these two agents develops. Naive cultures, which are sensitive to nanomolar concentrations of an enkephalin analog [Crain et al, 1978] become tolerant after exposure at 37°C for a 2- or 3-day period to low concentrations of this analog (10 nM). Tolerance is measured by the need to increase the opioid concentration necessary for effective bioelectric depression by 10-fold.

Iontophoretic application of endorphins to single nerve cells causes an inhibition of electrical activity. This observation is seen in most brain regions. The exceptions are the Renshaw cells of the spinal cord [Davies and Dray, 1978] and the pyramidal neurons within the hippocampus [Taylor et al, 1979]. In the case of the hippocampus, there is good evidence that the excitation results from disinhibition. All of these effects are naloxone reversible.

The effects of narcotic analgesics on pituitary hormone release are also observed with endorphins (see review by [Meites et al, 1979]). Like morphine, they stimulate the release of growth hormone, prolactin, and ACTH and inhibit the release of leuteinizing hormone, follicle stimulating hormone, and thyrotropin. All these effects are reversible by naloxone and this is, in fact, one of the rare instances where the administration of this antagonist to a naive animal produces a measurable effect. These effects are opposite to those seen when opiate agonists are administered. It is possible that one of the physiological roles of the endorphins may lie in the regulation of the anterior pituitary hormones. Accumulated evidence points to an action of the endorphins at the level of the hypothalamus rather than affecting directly the pituitary gland.

In contrast to the very short-lived analgesia produced by the enkephalins, analgesia produced by β-endorphin lasts for hours. The short duration of the action of enkephalins appears to be the result of their high sensitivity to degradation by peptidases. β-Endorphin, presumably due to the presence of consecutive lysine amino acids at positions 89 and 90, is less susceptible to enzymatic degradation [Geisow et al, 1977]. With the assumption that greater stability would lead to increased potency, a search was begun to find analogs of enkephalin resistant to the action of peptidases. Numerous analogs of enkephalin have been synthesized and tested for potency. The most significant finding was that substitution of D-alanine for the glycine in the 2-position resulted in both increased stability as well as increased potency and duration of action [Pert et al, 1976]. The increased stability is consistent with degradation by an aminopeptidase, which is further supported by the observed release of free tyrosine when enkephalins are incubated with rat and human plasma and rat brain homogenates [Hambrook et al, 1976]. Substitutions at position 5 further increase stability suggesting that carboxypeptidases are also involved in enkephalin breakdown [Bajusz et al, 1976]. A number of enkephalin analogs have been synthesized that are stable enough to exhibit potency after parenteral administration. Such a compound is FK 33-824 (Tyr-D-ala-Gly-MePhe-Met(O)-ol) which was synthesized by Sandoz, Ltd., and is a potent long-lasting analgesic [Roemer et al, 1977].

Biochemistry of Endorphins

How endorphins are synthesized and broken down, where they are stored, and how and when they are released are problems that investigators are still in the midst of solving. The observation that severe stress leads to simultaneous and proportional release into the circulatory system of both ACTH and β-endorphin was taken as an indication that they are coordinately regulated and concomitantly stored [Guillemin et al, 1977]. It was later shown that individual cells of the pituitary gland actually contain both of these peptides [Bloom et al,

1978]. In searching for a common precursor molecule, Mains et al [1977] found strong evidence that this precursor was a glycoprotein with a molecular weight of 31,000 daltons. They named this simply 31K protein, but the term suggested by S. Udenfriend, pro-opiocortin, is also used. This polypeptide consists of a 16,000 MW N-terminal glycoprotein of unknown function, followed by ACTH and finally LPH at the C-terminal end. A similar, but separate, biosynthetic system is present in bovine hypothalamus [Liotta et al, 1979].

The observations that neurons contain either, but never both, enkephalin or β-endorphin [Bloom et al, 1978] and that areas of the brain having high concentrations of enkephalin (such as the striatum and substantia nigra) exhibit no tract of β-endorphin or LPH [Rossier et al, 1977], coupled with the previously stated finding of the absence of an effect on enkephalin content in the brain following hypophysectomy, all pointed to a separate biosynthetic pathway for the enkephalins. Following intraventricular infusion of [^{3}H]amino acids, Yang et al [1978] isolated tritium labeled enkephalin. Tritium labeled enkephalin was also isolated from pig ileum myenteric plexus and from slices of striatum from guinea pig brain, following their incubation with [^{3}H]tyrosine [Hughes et al, 1978]. Finally, precursors or precursor fragments of Met- and Leu-enkephalin have been reported to be isolated from porcine hypothalami [Kangawa et al, 1979; Huang et al, 1979] and bovine adrenal medulla [Kimura et al, 1980]. These fragments are unrelated to porcine or bovine β-endorphin.

Putative Physiological Roles of Endorphins

The involvement of the endorphin-opiate receptor system in a variety of physiological, behavioral, and pathological states has been the subject of many reports. These studies have frequently been carried out by demonstrating that a given phenomenon is affected by specific opiate antagonist or endorphin administration. Using this protocol, endorphins have been implicated in memory, overeating, sexual activity, and some types of mental diseases. However, the best evidence for the involvement of the endorphin-opiate receptor system lies in the area of pain modulation. The distribution of opioid peptides and their receptors in the areas of the CNS involved with the conduction and modulation of pain and the demonstration of the analgesic potency of the endorphins strongly imply a role for this system in nociception.

If the assumption is made that the endorphins are endogenous analgesics, then the administration of opiate antagonists such as naloxone should lower pain thresholds or exacerbate pain. This expectation was not upheld in normal human subjects [El-Sobky et al, 1976; Grevert and Goldstein, 1978]. Perhaps these negative results are attributable to the experimental paradigms of pain induction, since naloxone administration to patients with postoperative pain caused hyperalgesia [Levine et al, 1978a]. Using the hot plate technique, Jacob et al [1974] were able to measure enhanced nociceptive responses in rats. The effect of naloxone on pain thresholds in animals is synchronous with the diurnal changes in levels of endogenous opiates [Frederickson et al, 1977].

Naloxone is capable of reversing the analgesia produced by means other than the administration of exogenous opiates. Thus, analgesia produced by electrical stimulation or acupuncture has been shown to be antagonized by naloxone in animals [Akil et al, 1976; Pomeranz and Chiu, 1976] and humans [Adams, 1976; Mayer et al, 1977]. Placebo produced analgesia has also been shown to be reversed by the administration of naloxone [Levine et al, 1978b].

Good success in the relief of intractable pain in patients by electrical stimulation via implanted electrodes in the periventricular gray region has been reported by Richardson and Akil [1977]. A multipolar electrode was implanted near the posterior aspect of the third ventricle, medial to the nucleus parafascicularis and in close proximity to the posterior comissure. The electrode was then connected to a Medtronics receiver implanted in the chest. A Medtronics stimulator allows patients to self-administer current at desired parameters (amplitude, frequency, etc). Significant or complete pain relief with periventricular stimulation was noted by patients. This analgesia was blocked by naloxone in 80% of the cases. The stimulation produced analgesia was effectual for a period of several months and in some cases even years. During surgery an intraventricular catheter was used to withdraw samples of ventricular fluid prior to and at intervals after electrical stimulation. RIAs were performed on these samples. While a less than twofold increase was seen in enkephalin-like immunoreactivity [Akil et al, 1978a], a 15-fold increase in β-endorphin-like immunoreactivity was observed [Akil et al, 1978b]. Thus, much evidence, albeit circumstantial, has been accumulated which points to the involvement of the endorphins originating in the brain in the modulation of pain.

Controversial evidence supporting speculations concerning a possible role of endorphins in the etiology of mental illness falls into two categories: (a) that severe psychiatric disorders may be caused by excess endorphin, (b) that endorphin level deficiency may lie at the root of mental illness. Both viewpoints are compatible with the hypothesis that variation from normal levels of endorphins may be related to psychiatric illness. The origins of these speculations began with the publication of back-to-back articles reporting that catatonic-like immobility, a component of certain types of schizophrenia, can be produced in the rat by intracerebral injection of small amounts of β-endorphin, in a naloxone reversible manner [Bloom et al, 1976; Jaquet and Marks, 1976]. The observation of elevated levels of β-endorphin-like material in the cerebrospinal fluid of schizophrenic individuals [Terenius et al, 1976] added further support to the above hypothesis. The clinical improvement in schizophrenic patients following repeated hemodialysis [Wagemaker and Cade, 1977], which was proposed to be due to the removal of leucine^5-b-endorphin [Palmour et al, 1979], suggests an endorphin excess in schizophrenia. However, the presence of leucine^5-β-endorphin in the hemofiltrates of schizophrenic and control patients could not be confirmed [Lewis et al, 1979]. If elevated levels of opioid peptides are related to schizophrenia, one would expect naloxone to have an effect on psychiatric symptoms. Out of 16 separate studies on the effect of opiate antagonists on schizophrenia (see [Gerner et al, 1980]), nine studies reported no change and seven studies reported dimunition of some symptoms, especially auditory hallucinations.

On the other side of the coin, evidence supporting an endorphin deficiency in schizophrenia, while being equally tenuous, has been accumulated. In three separate single-blind studies, psychiatric symptoms were decreased in schizophrenic patients when administered: the Met-enkephalin analog, FK 33-824 [Jorgensson et al, 1979], des-Tyr-β-endorphin [Verhoeven et al, 1979], or β-endorphin [Kline et al, 1977]. In two independent double-blind, crossover designed experiments, Berger et al [1980] and Gerner et al [1980] found no clinically obvious improvement in schizophrenic symptoms following β-endorphin administration. Gerner et al [1980], however, did find a significant improvement in depressed patients 2–4 hr after β-endorphin treatment. It is obvious that additional data collected under strictly controlled conditions are needed to, first, resolve all of these ambiguous findings and, second, find new support for a relationship between the endorphin-opiate receptor system and mental disease.

Many of the criteria for the classification, neurotransmitter, have been met by the enkephalins. They are located at nerve terminals [Watson et al, 1978], are readily degraded, and can be released by electrical stimulation, as in the guinea pig ileum [Puig et al, 1977] or by high concentrations of potassium ions, as was demonstrated in brain slices [Iversen et al, 1978]. The finding that endorphins inhibit the release of various neurotransmitters [Taube et al, 1976; Loh et al, 1977; Jhamandas et al, 1977], along with the demonstration that opiate receptors can have a presynaptic location, has given rise to the concept that the enkephalins and perhaps β-endorphin are neuromodulators, ie, the degree to which they are released modulates the release of classical neurotransmitters.

Finally, it must be emphasized strongly that definitive evidence for establishing a physiological role for any of the endorphins has, as yet, not been forthcoming. However, the fact that the endorphin-opiate receptor system is found in so many species along the evolutionary scale indicates plainly an important common functionality in the maintenance of the organism.

CONCLUDING REMARKS

This chapter has attempted to present a detailed overview of our present state of knowledge of opiate receptors and their endogenous ligands. The sequence of events in our acquisition of information in the opiate field is quite unique.

Most neuroactive substances were identified before their receptor sites were discovered and characterized. However, in the present case, establishment of the existence of the opiate receptors preceded by 2 years the discovery of their endogenous ligands. This general plan of investigation has been applied to the fairly recent demonstration of specific binding sites for benzodiazepine in the CNS. The search is well under way for its endogenous ligand, a possible tranquilizing neuroactive substance.

It is evident that because of limitations of space some important avenues of research have not been discussed; for instance, the work being carried out in attempts to elucidate the transducing portion of the receptor and the postbinding events triggered by it. Reports have appeared suggesting the involvement of adenylate cyclase and some evidence also exists pointing to a role of conformational changes in the opiate receptor and/or the surrounding membrane. Perhaps the work being carried out in opiate receptor solubilization and in studies on the heterogeneity of these receptors will clarify in the future the sequellae of events following the binding of opiates and endorphins to their receptors.

ACKNOWLEDGMENTS

The research, carried out by a team headed by Dr. Eric J. Simon, was supported by Grant DA-00017 from the National Institute on Drug Abuse. We also gratefully acknowledge financial assistance from Hoffmann–LaRoche, Inc., and E.I. Dupont de Nemours & Company.

REFERENCES

1. Adams JE: Naloxone reversal of analgesia produced by brain stimulation in the human. Pain 2:161–166, 1976.
2. Akil H, Mayer DJ, Liebeskind JC: Antagonism of stimulation-produced analgesia by naloxone, a narcotic antagonist. Science 191:961–962, 1976.
3. Akil H, Richardson DE, Hughes J, Barchas JD: Enkephalin-like material elevated in ventricular cerebrospinal fluid of pain patients after analgetic focal stimulation. Science 201:463–465, 1978a.
4. Akil H, Richardson DE, Barchas JD, Li CH: Appearance of β-endorphin-like immunoreactivity in human ventricular cerebrospinal fluid upon analgesic electrical stimulation. Proc Nat Acad Sci USA 75:5170–5172, 1978b.
5. Alumets J, Hakanson R, Sundler F, Thorell J: Neuronal localization of immunoreactive enkephalin and β-endorphin in the earthworm. Nature (London) 279:805–806, 1979.
6. Atweh SF, Kuhar MJ: Autoradiographic localization of opiate receptors in rat brain. I. Spinal cord and lower medulla. Brain Res 124:53–67, 1977a.
7. Atweh SF, Kuhar MJ: Autoradiographic localization of opiate receptors in rat brain. III. The telencephalon. Brain Res 134:393–406, 1977b.
8. Atweh SF, Kuhar MJ: Autoradiographic localization of opiate receptors in rat brain. II. The brainstem. Brain Res 129:1–12, 1977c.
9. Bajusz S, Rónai AZ, Székely JI, Dunai-Kóvács S, Berzétei I, Graf L: Enkephalin analogs with enhanced opiate activity. Acta Biochem Biophys Acad Sci Hung 11:305–309, 1976.
10. Belluzzi JD, Grant N, Garsky V, Sarantakes D, Wise CD, Stein L: Analgesia induced in vivo by central administration of enkephalin in rat. Nature (London) 260:625–626, 1976.

11. Berger PA, Watson SJ, Akil H, Elliot GR, Rubin RT, Pfefferbaum A, Davis KL, Barchas JD, Li CH: β-Endorphin and schizophrenia. Arch Gen Psychiat 37:635–640, 1980.
12. Bidlack JM, Abood LG: Solubilization of the opiate receptor. Life Sci 27:331–340, 1980.
13. Bloom F, Segal D, Ling N, Guillemin R: Endorphins: Profound behavior effects in rats suggest new etiological factors in mental illness. Science 194:630–632, 1976.
14. Bloom F, Battenberg ELF, Rossier J, Ling N, Leppaluoto J, Vargo TM, Guillemin R: Endorphins are located in the intermediate and anterior lobes of the pituitary gland, not in the neurohypophysis. Life Sci 20:43–48, 1977.
15. Bloom F, Battenberg E, Rossier J, Ling N, Guillemin R: Neurons containing β-endorphin in rat brain exist separately from those containing enkephalin: Immunocytochemical studies. Proc Nat Acad Sci 75:1591–1595, 1978.
16. Bradbury AF, Smyth DG, Snell CR, Birdsall NJM, Hulme EC: C fragment of lipotropin has a high affinity for brain opiate receptors. Nature (London) 260:793–795, 1976.
17. Bradbury A, Smyth D, Snell C, Deakin J, Wendlant S: Comparison of the analgesic properties of lipotropin C-fragment and stabilized enkephalins in the rat. Biochem Biophys Res Commun 74:748–754, 1977.
18. Bruni JF, Watkins WB, Yen SSC: β-Endorphin in the human pancreas. J Clin Endocrin Metabol 49:649–651, 1979.
19. Buscher HH, Hill RC, Römer D, Cardinaux F, Closse A, Hauser D, Pless J: Evidence for analgesic activity of enkephalin in the mouse. Nature (London) 261:423–425, 1976.
20. Caron MG, Lefkowitz RJ: Solubilization and characterization of the β-andrenergic receptor binding sites of frog erythrocytes. J Biol Chem 251:2374–2384, 1976.
21. Cheung A, Goldstein A: Failure of hypophysectomy to alter brain content of opioid peptides (endorphins). Life Sci 19:1005–1008, 1976.
22. Crain SM, Peterson ER: Enhanced afferent synaptic function in fetal mouse spinal cord-sensory ganglion explants following NGF-induced ganglion hypertrophy. Brain Res 79:145–152, 1974.
23. Crain SM, Peterson ER: Development of specific sensory-evoked synaptic networks in fetal mouse cord-brainstem cultures. Science 188:275–278, 1975.
24. Crain SM, Peterson ER, Crain B, Simon EJ: Selective opiate depression of sensory-evoked synaptic networks in dorsal horn regions of spinal cord cultures. Brain Res 133:162–166, 1977.
25. Crain SM, Crain B, Peterson ER, Simon EJ: Selective depression by opioid peptides of sensory-evoked dorsal-horn network responses in organized spinal cord cultures. Brain Res 157:196–201, 1978.
26. Crain SM, Crain B, Finnigan T, Simon EJ: Development of tolerance to opiates and opioid peptides in cultures of mouse spinal cord with attached dorsal root ganglia. Life Sci 25:1797–1802, 1977.
27. Cox BM, Opheim KE, Teschemacher H, Goldstein A: A peptide-like substance from pituitary that acts like morphine. 2. Purification and properties. Life Sci 16:1777–1782, 1975.
28. Cox BM, Goldstein A, Li CH: Opioid activity of a peptide, β-lipotropin-(61-91) derived from β-lipotropin. Proc Nat Acad Sci 73:1821–1823, 1976.
29. Creese I, Snyder SH: Receptor binding and pharmacological activity of opiates in the guinea pig intestine. J Pharmacol Exp Ther 194:205–219, 1975.
30. Davies J, Dray A: Pharmacological and electrophysiological studies of morphine and enkephalin in rat supraspinal neurones and cat spinal neurones. Br J Pharmacol 63:87–96, 1978.
31. DiGuilio AM, Yang H-Y, Lutold B, Fratta W, Hong J, Costa E: Characterization of enkephalin-like material extracted from sympathetic ganglia. Neuropharmacology 17:989–992, 1978.
32. Elde R, Hökfelt T, Johansson O, Terenius L: Immunohistochemical studies using antibodies to leucine enphalin: Initial observations on the nervous system of the cat. Neuroscience 1:349–351, 1976.

33. El-Sobky A, Dostrovsky JO, Wall PD: Lack of effect of naloxone on pain perception in humans. Nature (London) 263:783–785, 1976.
34. Florez J, Mediavilla A: Respiratory and cardiovascular effect of met-enkephalin applied to the ventral surface of the brain stem. Brain Res 183:585–590, 1977.
35. Frederickson RCA, Burgis V, Edwards JD: Hyperalgesia induced by naloxone follows diurnal rhythm in responsivity to painful stimuli. Science 198:756–758, 1977.
36. Geison M, Deakin J, Dostrovsky J, Smyth D: Analgesic activity of lipotropin C-fragment depends on carboxyl terminal tetrapeptide. Nature (London) 269:167–168, 1977.
37. Gerner RH, Catlin DH, Gorelick DA, Hui KK, Li CH: β-Endorphin: Intravenous infusion causes behavioral change in psychiatric inpatients. Arch Gen Psychiat 37:642–647, 1980.
38. Gilbert PE, Martin WR: The effects of morphine- and nalorphine-like drugs in the nondependent, morphine-dependent and cyclazocine-dependent chronic spinal dog. J Pharmacol Exp Therap 198:66–82, 1976.
39. Glaser T, Van Calker D, Hubner K, Stadtkus C, Hamprecht B: Presence of opioid peptides in a neuroblastoma x glioma hybrid cell line. Eur J Pharmacol 65:319–320, 1980.
40. Goldstein A, Lowney LI, Pal BK: Stereospecific and nonspecific interactions of the morphine congener levorphanol in subcellular fractions of mouse brain. Proc Nat Acad Sci 68:1742–1747, 1971.
41. Goldstein A, Tachibana S, Lowney LI, Huckapiller M, Hood L: Dynorphin (1–13), an extraordinarily potent opioid peptide. Proc Nat Acad Sci 76:6666–6676, 1979.
42. Grevert P, Goldstein A: Endorphins: Naloxone fails to alter experimental pain or mood in humans. Science 199:1093–1095, 1978.
43. Guillemin R, Ling N, Burgus R: Endorphines, peptides, d'origine hypothalamique et neurohypophysaire à actité morphinomimetique. Isolement et structure moleculaire de l'endorphine. C R Acad Sci (Paris) 282:783–785, 1976.
44. Guillemin R, Vargo T, Rossier J, Minick S, Ling N, Rivier C, Bloom F: β-Endorphin and adrenocorticotropin are secreted concomitantly by the pituitary. Science 197:1367–1369, 1977.
45. Hambrook J, Morgan B, Rance K, Smith C: Mode of deactivation of the enkephalins by rat and human plasma and rat brain homogenates. Nature (London) 262:782–783, 1976.
46. Hiller JM, Pearson J, Simon EJ: Distribution of stereospecific binding of the potent narcotic analgesic etorphine in the human brain: Predominance in the limbic system. Res Commun Chem Pathol Pharmacol 6:1052–1062, 1973.
47. Hiller JM, Simon EJ, Crain SM, Peterson ER: Opiate receptors in cultures of fetal mouse dorsal root ganglia (DRG) and spinal cord: Predominance in DRG neurites. Brain Res 145:396–400, 1978a.
48. Hiller JM, Simon EJ, Crain SM, Peterson ER: Opiate receptor distribution in organized cultures of fetal mouse spinal cord and dorsal root ganglia. In Van Ree JM and Terenius L (eds): "Characteristics and Function of Opioids." Amsterdam: Elsevier–North Holland Biomed. Press, 1978b, pp 477–478.
49. Hökfelt T, Elde R, Johansson O, Terenius L, Stein L: The distribution of enkephalin immunoreactive cell bodies in the rat central nervous system. Neurosci Lett 5:25–31, 1977.
50. Holaday JW, Wei E, Loh HH, Li CH: Endorphins may function in heat adaptation. Proc Nat Acad Sci 75:2923–2927, 1978.
51. Houck JC, Kimball C, Chang C, Pedigo NW, Yamamura HI: Placental β-endorphin-like peptides. Science 207:78–80, 1980.
52. Huang WY, Chang RCC, Kastin AJ, Cox DH, Schally AV: Isolation and structure of promethionine-enkephalin: Potential enkephalin precursor from porcine hypothalamus. Proc Nat Acad Sci 76:6177–6180, 1979.
53. Hughes J: Isolation of an endogenous compound from the brain with properties similar to morphine. Brain Res 88:295–308, 1975a.

54. Hughes J: Search for the endogenous ligand of the opiate receptor. Neurosci Bull 13:55–58, 1975b.
55. Hughes J, Kosterlitz HW, Leslie FM: Effect of morphine on adrenergic transmission in the mouse vas deferens. Assessment of agonist and antagonist potencies of narcotic analogues. Br J Pharmacol 53:371–381, 1975a.
56. Hughes J, Smith TW, Kosterlitz HW, Fothergill LA, Morgan BA, Morris HR: Identification of two related pentapeptides from the brain with potent opiate agonist activity. Nature (London) 258:577–579, 1975b.
57. Hughes J, Kosterlitz HW, McKnight AT: The incorporation of ^{3}H-tyrosine into the enkephalins of striatal slices of guinea-pig brain. Br J Pharmacol 63:396P, 1978.
58. Hitzeman RJ, Hitzeman BA, Loh HH: Binding of ^{3}H-naloxone in the mouse brain: Effect of ions and tolerance development. Life Sci 14:2393–2404, 1974.
59. Ingoglia NA, Dole VP: Localization of *d*- and *l*-methadone after intraventricular injection into rat brains. J Pharmacol Exp Ther 175:84–87, 1970.
60. Iversen LL, Iversen SD, Bloom FE, Vargo T, Guillemin R: Release of enkephalin from rat globus pallidus in vitro. Nature (London) 271:679–681, 1978.
61. Jacob JJ, Tremblay EC, Colombel M-C: Facilitation de réactions nociceptives par la naloxone chez la souris et chez rat. Psychopharmacology 37:217–223, 1974.
62. Jacquet YF, Marks N: The C-fragment of β-lipotropin: An endogenous neuroleptic or antipsychotogen? Science 194:632–635, 1976.
63. Jhamandas K, Sawynok J, Sutak M: Enkephalin effects on release of brain acetylcholine. Nature (London) 269:433–434, 1977.
64. Jorgensson A, Fog R, Veilis B: Synthetic enkephalin analogue in treatment of schizophrenia. Lancet 1:935, 1979.
65. Kangawa K, Matsuo H, Igarashi I: α-Neo-endorphin: A "big" leu-enkephalin with potent opiate activity from porcine hypothalami. Biochem Biophys Res Commun 86:153–160, 1978.
66. Kastin A, Scollan E, King M, Schally A, Coy D: Enkephalin and a potent analog facilitate maze performance after intraperitoneal administration in rats. Pharmac Biochem Behav 5:691–695, 1976.
67. Kastin AJ, Scollan EL, Ehrensing RH, Schally AV, Coy DH: Enkephalin and other peptides reduce passiveness. Pharmacol Biochem Behavior 9:515–519, 1978.
68. Katz RJ, Roth KA: Stress induced grooming in the rat—an endorphin mediated syndrome. Neurosci Lett 13:209–212, 1979.
69. Kimura S, Lewis RV, Stern AS, Rossier J, Stein S, Udenfriend S: Probable precursors of (leu)enkephalin and (met)enkephalin in adrenal medulla: Peptide of 3-5 kilodaltons. Proc Nat Acad Sci 77:1681–1685, 1980.
70. Kline N, Li C, Lehmann J, Lajtha A, Laski E, Cooper T: β-Endorphin-induced changes in schizophrenic and depressed patients. Arch Gen Psychiat 34:1111–1113, 1977.
71. Kosterlitz HW, Waterfield AA: In vitro models in the structure-activity relationship of narcotic analgesics. Ann Rev Pharmacol Toxicol 15:29–47, 1975.
72. Kuhar MJ, Pert CB, Snyder SH: Regional distribution of opiate receptor binding in monkey and human brain. Nature (London) 245:447–450, 1973.
73. LaMotte C, Pert CB, Snyder SH: Opiate receptor binding in primate spinal cord: Distribution and changes after dorsal root section. Brain Res 112:407–412, 1976.
74. Lewis RV, Gerber LD, Stein S, Stephen RL, Grosser BI, Velick SF, Udenfriend S: On β_H-Leu^5-endorphin and schizophrenia. Arch Gen Psychiat 36:237–239, 1979.
75. Levine JD, Gordon NC, Jones RT, Fields HL: The narcotic antagonist naloxone enhances clinical pain. Nature (London) 272:826–827, 1978a.
76. Levine JD, Gordon NC, Field HL: The mechanism of placebo analgesia. Lancet 2:654–657, 1978b.

77. Lord JAH, Waterfield AA, Hughes J, Kosterlitz HW: Endogenous opioid peptides: Multiple agonists and receptors. Nature (London) 267:495–499, 1977.
78. Li CH: Lipotropin, a new active peptide from pituitary glands. Nature (London) 201:924, 1964.
79. Li CH, Chung D: Isolation and structure of an untriakontapeptide with opiate activity from camel pituitary glands. Proc Nat Acad Sci 73:1145–1148, 1976.
80. Liebeskind JC, Mayer DJ, Akil H: Central mechanisms of pain inhibition: Studies of analgesia from focal brain stimulation. In Bonica JJ (ed): "Advances in Neurology." Vol. 4, International Symposium on Pain. New York: Raven Press, 1974, pp 261–268.
81. Linnoila RE, Diaugustine RP, Miller RJ, Chang KJ, Cautrecasas P: An immunohistochemical and radioimmunological study of the distribution of (Met^5)- and (Leu^5)-enkephalin in the gastrointestinal tract. Neuroscience 3:1187–1196, 1978.
82. Liotta AS, Gildersleeve D, Brownstein MJ, Krieger DT: Biosynthesis *in vitro* of immunoreactive 31,000 dalton corticotropin/β-endorphin-like material by bovine hypothalamus. Proc Nat Acad Sci 76:1448–1452, 1979.
83. Loh H, Brase D, Sampath-Khanna S, Mar J, Li C: β-Endorphin in vitro inhibition of striatal dopamine release. Nature (London) 264:567–568, 1976.
84. Mains RE, Eipper BA, Ling N: Common precursor to corticotropins and endorphins. Proc Nat Acad Sci 74:3014–3018, 1977.
85. Mayer DJ, Liebeskind JC: Pain reduction by focal electrical stimulation of the brain: An anatomical and behavioral analysis. Brain Res 68:73–93, 1974.
86. Mayer DJ, Hayes R: Stimulation-produced analgesia: Development of tolerance and cross-tolerance to morphine. Science 188:941–943, 1975.
87. Mayer DJ, Price DD, Rafii A: Antagonism of acupuncture analgesia in man by the narcotic antagonist naloxone. Brain Res 121:368–372, 1977.
88. Meek JL, Bohan TP: Use of high pressure liquid chromatography to study enkaphalins. In Costa E and Trabucchi M (eds): "The Endorphins: Advances in Biochemical Psychopharmacology." Vol. 18, New York: Raven Press, 1978, pp 141–148.
89. Meites J, Bruni JF, Van Vugt DA, Smith AF: Relation of endogenous opioid peptides and morphine to neuroendocrine functions. Life Sci 24:1325–1336, 1979.
90. Mello NK, Mendelson JH: Self-administration of an enkephalin analog by rhesus monkey. Pharmac Biochem Behav 9:579–586, 1978.
91. Palmour R, Ervin F, Wagemaker H, Cade R: Characterization of a peptide from the serum of psychotic patients. In Usdin E, Bunney WE, and Kline NS (eds): "Endorphins in mental health research." London: Macmillan Press, 1979, pp 581–593.
92. Pasternak GW, Snyder SH: Opiate receptor binding: Effects of enzymatic treatment. Mol Pharmacol 10:183–193, 1973.
93. Pasternak GW, Goodman R, Snyder SH: An endogenous morphine-like factor in mammalian brain. Life Sci 16:1765–1769, 1975.
94. Paton WDM: The action of morphine and related substances on contraction and on acetylcholine output of coaxially stimulated guinea pig ileum. Br J Pharmacol Chemother 12:119–127, 1957.
95. Pert CB, Snyder SH: Opiate receptor: Demonstration in nervous tissue. Science 179:1011–1014, 1973.
96. Pert CB, Snyder SH: Opiate receptor binding of agonists and antagonists affected differentially by sodium. Mol Pharmacol 10:868–879, 1974.
97. Pert CB, Snyder SH: Differential interactions of agonists and antagonists with the opiate receptor. Neurosci Res Prog Bull 13:73–79, 1975.
98. Pert CB, Aposhian D, Snyder SH: Phylogenetic distribution of opiate receptor binding. Brain Res 75:356–361, 1974a.
99. Pert CB, Snowman AM, Snyder SH: Localization of opiate receptor binding in synaptic membranes of rat brain. Brain Res 70:184–188, 1974b.

100. Pert CB, Pert A, Chang J-K, Fong BTW: (D-Ala^2)-Met-enkephalinamide: A potent, long-lasting synthetic pentapeptide analgesic. Science 194:330–332, 1976.
101. Pert CB, Taylor DB: Type 1 and Type 2 opiate receptors: A subclassification scheme based upon GTP's differential effects on binding. In Way EL (ed): "Endogenous and Exogenous Opiate Agonists and Antagonists." New York: Pergamon Press, 1979, pp 87–90.
102. Pfefferbaum A, Berger PA, Elliot GR, Tinkleberg JR, Kopell BS, Barchas JD, Li CH: Human EEG response to beta-endorphin. Psychiat Res 1:83–88, 1979.
103. Polak JM, Sullivan SN, Bloom SR, Facer R, Pearse AGE: Enkephalin-like immunoreactivity in the human gastrointestinal tract. Lancet 1:972–974, 1977.
104. Pomeranz B, Chiu D: Naloxone blockade of acupuncture analgesia: Endorphin implicated. Life Sci 19:1757–1762, 1976.
105. Puig MM, Gascon P, Craviso GL, Musacchio JM: Endogenous opiate ligand: Electrically induced release in the guinea pig ileum. Eur J Pharmacol 45:205–206, 1977.
106. Richardson DE, Akil H: Pain reduction by electrical brain stimulation in man. Part 2: Chronic self-administration in the periventricular gray matter. J Neurosurg 47:184–194, 1977.
107. Robson LE, Kosterlitz HW: Specific protection of the binding sites of D-ala^2-D-Leu^5-enkephalin (δ-receptors) and dihydromorphine (μ-receptors). Proc R Soc London (Biol) 205:425–432, 1979.
108. Roemer D, Buescher HJ, Hill RC, Pless J, Bauer W, Cardinaux F, Closse A, Hauser D, Huguenin R: A synthetic enkephalin analogue with prolonged parenteral and oral analgesic activity. Nature (London) 268:547–549, 1977.
109. Rossier J, Bayon A, Vargo TM, Ling N, Guillemin R, Bloom F: Radioimmunoassay of brain peptides. Evaluation of a methodology for the assay of β-endorphin and enkaphalin. Life Sci 21:847–852, 1977.
110. Rubinstein M, Stein S, Gerber LD, Udenfriend S: Isolation and characterization of opioid peptides from rat pituitary: β-Lipotropin. Proc Nat Acad Sci 74:3052–3055, 1977.
111. Ruegg UT, Hiller JM, Simon EJ: Solubilization of an active opiate receptor from *Bufo marinus*. Eur J Pharmacol 64:367–368, 1980.
112. Schultzberg M, Lundberg JM, Hökfelt T, Terenius L, Brandt J, Elde RP, Goldstein M: Enkephalin-like immunoreactivity in gland cells and nerve terminals of the adrenal medulla. Neuroscience 3:1169–1186, 1978.
113. Simantov R, Goodman R, Aposhian D, Snyder SH: Phylogenetic distribution of a morphine-like peptide "enkephalin." Brain Res 111:204–211, 1976.
114. Simantov L, Kuhar MJ, Uhl G, Snyder SH: Opioid peptide enkephalin immunohistochemical mapping in the rat CNS. Proc Nat Acad Sci 74:467–471, 1977.
115. Simon EJ: Morphine and related drugs. In Jacoby WB and Wilchek M (eds): "Affinity Techniques: Methods in Enzymology." Vol. 34B, New York: Academic Press, 1974, pp 619–623.
116. Simon EJ: Opiate receptor binding with ^{3}H-etorphine. Neurosci Res Program Bull 13:43–50, 1975.
117. Simon EJ, Groth J: Kinetics of opiate receptor inactivation by sulfhydryl reagents: Evidence for conformational change in presence of sodium ions. Proc Nat Acad Sci 72:2404–2407, 1975.
118. Simon EJ, Hiller JM, Edelman I: Stereospecific binding of the potent narcotic analgesic ^{3}H-etorphine to rat brain homogenate. Proc Nat Acad Sci 70:1947–1949, 1973.
119. Simon EJ, Hiller JM, Groth J, Edelman I: Further properties of stereospecific opiate binding sites in rat brain: On the nature of the sodium effect. J Pharmacol Exp Ther 912:531–537, 1975a.
120. Simon EJ, Hiller JM, Edelman I: Solubilization of a stereospecific opiate-macromolecular complex from rat brain. Science 190:389–390, 1975b.

121. Simon EJ, Bonnet KA, Crain SM, Groth J, Hiller JM, Smith JR: Recent studies on the interaction between opioid peptides and their receptors. In Costa E and Trabucchi M (eds): "Advances in Biochemistry and Psychopharmacology." 22:335–346, 1980.
122. Simonds WF, Koski G, Streaty RA, Hjelmeland LM, Klee WA: Solubilization of active opiate receptors. Proc Nat Acad Sci 77:4623–4627, 1980.
123. Smith JR, Simon EJ: Selective protection of stereospecific enkephalin and opiate binding against inactivation by N-ethylmaleimide: Evidence for two classes of receptors. Proc Nat Acad Sci 77:281–284, 1980.
124. Stahl KD, Van Bever W, Janssen P, Simon EJ: Receptor affinity and pharmacological potency of a series of narcotic analgesic, antidiarrheal and neuroleptic drugs. Eur J Pharmacol 46:199–205, 1977.
125. Stefano GB, Kream RM, Zukin RS: Demonstration of stereospecific binding in the nervous tissue of the marine mollusc *Mytilus edulis*. Brain Res 181:440–445, 1980.
126. Strauss WL, Ghai G, Fraser CM, Venter JC: Detergent solubilization of mammalian cardiac and hepatic β-adrenergic receptors. Arch Biochem Biophys 196:566–573, 1979.
127. Taube HD, Borowski E, Endo T, Starke K: Enkephalin a potential modulator of noradrenaline release in rat brain. Acta Pharmacol Toxicol 33:377–384, 1976.
128. Taylor D, Hoffer B, Ziegelgänsberger W, Siggins G, Ling N, Seiger Å, Olson L: Opioid peptides excite pyramidal neurons and evoke epileptiform activity in hippocampal transplants in oculo. Brain Res 176:135–142, 1979.
129. Terenius L: Stereospecific interaction between narcotic analgesics and a synaptic plasma membrane fraction of rat cerebral cortex. Acta Pharmacol Toxicol 32:317–320, 1973a.
130. Terenius L: Characteristics of the "receptor" for narcotic analgesics in synaptic plasma membrane fraction from rat brain. Acta Pharmacol Toxicol 33:377–384, 1973b.
131. Terenius L, Wablström A: Inhibitor(s) of narcotic receptor binding in brain extracts and cerebrospinal fluid. Acta Pharmacol Toxicol 35, Suppl. 1:55, 1974.
132. Terenius L, Wahlström A: Search for an endogenous ligand for the opiate receptor. Acta Physiol Scand 94:74–81, 1975.
133. Terenius L, Wahlström A, Lindstrom L, Widerlov E: Increased CSF levels of endorphins in chronic psychosis. Neurosci Lett 3:157–162, 1976.
134. Teschemacher H, Opheim KE, Cox BM, Goldstein A: A peptide-like substance from pituitary that acts like morphine. 1. Isolation. Life Sci 16:1771–1776, 1975.
135. Tseng LF, Loh HH, Li CH: β-Endorphin as a potent analgesic by intravenous injection. Nature (London) 263:239–240, 1976.
136. Tseng LF, Loh HH, Li CH: β_h-endorphin: Antidiuretic effect in rats. Int J Peptide Protein Res 12:173–176, 1978.
137. Vauquelin G, Geynet P, Hanoune J, Strosberg AD: Isolation of adenylatecyclase-free, β-adrenergic receptor from turkey erythrocyte membranes by affinity chromatography. Proc Nat Acad Sci 74:3710–3714, 1977.
138. Verhoeven WMA, Van Praag HM, Van Ree JM, de Wied D: Improvement of schizophrenic patients treated with (Des-Tyr^1)-γ-endorphin (DT$_\gamma$E). Arch Gen Psychiatry 36:294–298, 1979.
139. Wagemaker H, Cade R: The use of hemodialysis in chronic schizophrenia. Am J Psychiatry 134:684–685, 1977.
140. Waterfield AA, Hughes J, Kosterlitz HW: Cross tolerance between morphine and methionine enkephalin. Nature (London) 260:624–625, 1976.
141. Watson SJ, Akil H, Sullivan SO, Barchas JD: Immunocytochemical localization of methionine enkephalin: Preliminary observations. Life Sci 21:733–738, 1977a.
142. Watson SJ, Barchas JD, Li CH: β-Lipotropin: Localization of cells and axons in rat brain by immunochemistry. Proc Nat Acad Sci 74:5155–5158, 1977b.

143. Watson SJ, Akil H, Richard III CW, Barchas JD: Evidence for two separate opiate peptide neuronal systems. Nature (London) 275:226–228, 1978.
144. Wei E, Loh H: Physical dependence of opiate-like peptides. Science 193:1262–1263, 1976.
145. Wilson RS, Rogers ME, Pert CB, Snyder SH: Homologous N-alkylnorketobemidones. Correlation of receptor binding with analgesic potency. J Med Chem 18:240–242, 1975.
146. Yang H-Y, Hong HS, Costa E: Regional distribution of leu and met enkephalin in rat brain. Neuropharmacology 16:303–307, 1977.
147. Yang H-YT, Hong HS, Fratta W, Costa E: Rat brain enkephalins: Distribution and biosynthesis. In Costa E and Trabucchi M (eds): "The Endorphins: Advances in Biochemistry and Psychopharmacology." Vol 18, New York: Raven Press, 1978, pp 149–160.
148. Zipser B: Identification of specific leech neurones immunoreactive to enkephalin. Nature (London) 283:857–858, 1980.

The Role of Peptides and Amino Acids as Neurotransmitters, pages 53-97
©1981 Alan R. Liss, Inc., 150 Fifth Avenue, New York, NY 10011

Insights on Function: Metabolism and Pharmacology of Taurine in the Brain

Ryan J. Huxtable

TAURINE: BACKGROUND AND PHYSICOCHEMICAL PROPERTIES

Taurine, or 2-aminoethane sulfonic acid, is an unusual amino acid. It is unusual in being a β-amino sulfonic acid instead of an α-amino carboxylic acid, for being chemically and metabolically stable, for not being a constituent of any protein, for occurring in mammalian tissues in high concentrations, and, despite this, for being little understood or known.

Taurine, $C_2H_7NO_3S$, is a simple substance of molecular weight 125.1. Its structure, with the structures of related amino acids, is shown on Figure 1. The sulfonic acid function is highly acidic, having a pK_a of 1.5. The amino group has a pK_b of 8.82. Its carboxylic analog, β-alanine (Fig. 1), has a pK_a of 3.55 and a pK_b of 10.24. Both functions of taurine are thus more acidic, and the compound is zwitterionic and hydrophilic. Inasmuch as the biological properties of a compound are an ineluctable consequence of its physicochemical properties, some relevant constants for a number of amino acids are given in Table I.

What do we know of taurine? Does it have any function in biological systems?

For many people, taurine represents an inert end product of sulfur metabolism, drifting uselessly through a biochemical sea until it runs afoul of some island of bacterial metabolism. The reason for thinking this is readily comprehended. Taurine is a difficult and somewhat mysterious substance. It is easier to decide that if it is hard to find a purpose for a compound then it has none, than it is to become involved in the study of a problem not readily tractable. However, taurine is now floating on a rising tide of interest, as can be seen in Figure 2. Figure 2A depicts the annual volume of literature on taurine as measured by two

Abbreviations used in text: μA, microamp; CA, cysteic acid; CNS, central nervous system; CSA, cysteine sulfinic acid; GABA, γ-aminobutyric acid; ICV, intracerebroventricular(ly); ISA, isethionic acid; PAPS, phosphoadenosine phosphosulfate.

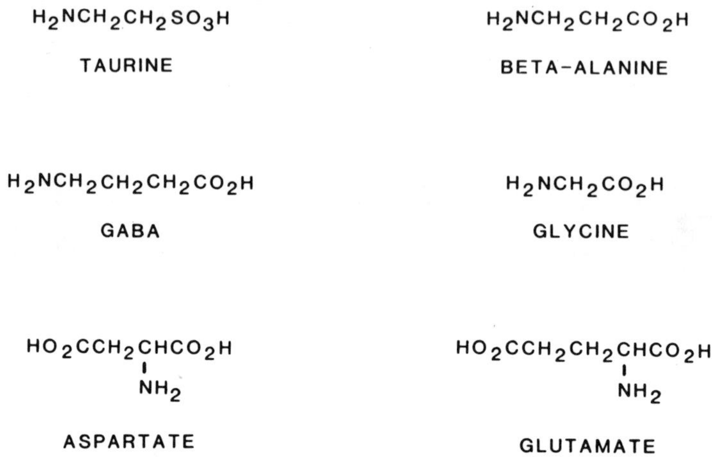

Fig. 1. Neuroactive amino acids. Acidic amino acids have excitatory actions on neurons, whereas αω-amino acids have inhibitory actions.

TABLE I. Physical Constants of Neuroactive Amino Acids

Amino acids	Solubility[a] (g/100 ml in water) (25°C)	(100°C)	Ionization constants[b] (pK_1)	(pK_2)	(pK_3)	Isoelectric point
TAU	10.48		1.5	8.82		5.16
GLY	24.99	67.17	2.34	9.60		5.97
ASP	0.5	6.9	1.88	3.65	9.60	2.77
GLU	0.86	14.0	2.16	4.32	9.96	3.24
β-ALA	freely sol.		3.55	10.24		6.89
GABA	freely sol.		4.03	10.56		7.30
ASN	2.98	55.1	2.02	8.80		5.41
GLN	3.6		2.17	9.13		5.65

[a]Data from Jacobsen and Smith [1968]; Windholz et al [1976]; and Jakubke and Jeschkeit [1977].
[b]Data from Johnston [1975]; Windholz et al [1976]; and Bender [1975].

rather arbitrary indices: cross-references to taurine in Science Citation Index, and titles containing taurine listed in Biological Abstracts. These indices ignore the considerable literature on the subject in which the word "taurine" is not part of the title. However, the relative increase in interest is obvious. Although not a pharmacologist by training, I have picked up enough habits from my colleagues to wish to show my data on a semi-log plot, as in a dose–response curve. The information in Figure 2A is reshown in Figure 2B on such a plot. Linear regres-

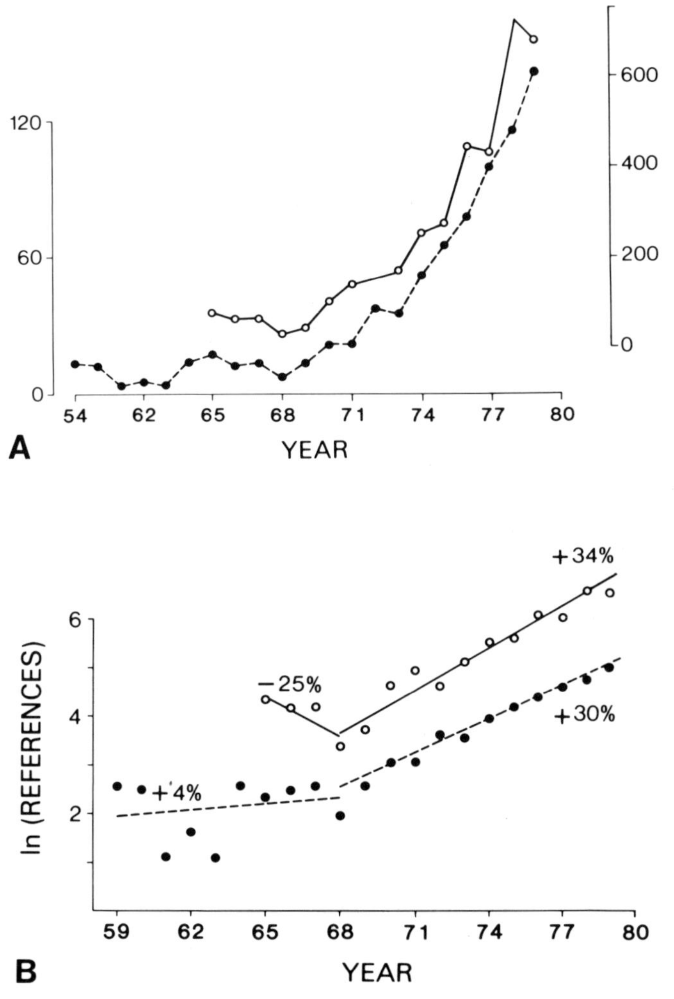

Fig. 2. (A) The taurine literature. Closed circles (left axis) indicate the number of papers appearing each year with the word "taurine" in their title (taken from Biological Abstracts). Open circles (right axis) gives the number of cross references to taurine in Science Citation Index each year. (B) This shows the same data as (A), plotted semilogarithmically. The percentages are the percent annual increases for the periods until 1968, and 1969–1979.

sion analysis indicates that between 1968 and 1979 literature on taurine grew at an annual rate of between 30 and 34%. Much of this interest has been fueled by considerations of the involvement of taurine in central nervous system function.

Despite the vast literature that has accumulated, our knowledge of taurine is limited and lacks integration. From a distance, the picture, although without detail, appears continuous, and it is possible to make general statements and to draw some inferences. As an area is highlighted, however, and scrutinized more closely, the apparent continuity dissolves like a magnified photograph, and what is seen are flecks of fact isolated in ignorance. I shall attempt in this article not to review the area in exhaustive detail, which would be laborious in the doing and unreadable in the result, but to point out, in a few chosen areas, where the deficiencies in knowledge lie and the importance of remedying these deficiencies if we are to obtain a balanced view of the involvement of taurine in CNS phenomena.

Taurine and the Evolutionary Tree

Apart from its occurrence in certain fungi and the pollen of a few plants, taurine is absent from the plant kingdom and is limited to the animal and microbial kingdoms [Close, 1960; Marquardt and Vogg, 1952]. It is an ancient compound, being found in single-celled algae [Ericson and Carlson, 1954]. It is a general rule in biochemistry, if a teleological construction may be used, that the compounds are made first and then functions are found for them. Nature, like a lazy chemist, tends to look on the shelf for a compound that will do rather than synthesize a new substance. Such is the case with the catecholamines, for example, which are found in a wide variety of plants and which were present in biomass before their indispensability was discovered by central nervous systems. The same is true for taurine. As it has progressed up the phylogenetic tree, new functions have accreted to it. In marine invertebrates and various fish, it appears to serve the function of an osmotic agent [Lange, 1963, 1964; Lange and Fugelli, 1965; Hoyaux et al, 1976]. When such species are moved from a marine to a fresh water environment, taurine concentrations fall, and vice versa [Fugelli and Zachariassen, 1976]. Taurine also appears to be involved in osmotic regulation of mammalian brain [Thurston et al, 1980]. The action of taurine on mammalian drinking behavior may be derived from the osmotic effect of taurine [Hruska et al, 1975]. In the annelid worms, a phosphorylated derivative of taurine, N-phosphorylguanidinoethyl sulfonate, serves as a secondary phosphagen, analogous to the role that phosphoryl creatine serves in mammalian muscle [Ennor and Morrison, 1958]. By the time we reach the sharks, fishes which hitherto have conjugated sulfate and cholic acid to give surfactive agents necessary for the digestion of fats, now make bile acid from taurine and cholic acid. Some mammals still do this; others such as the rabbit and guinea pig no longer produce taurocholic acid [Haslewood, 1962], although they apparently have the necessary enzymatic machinery to do so [Vessey, 1977].

Most animals, be they arthropods or mammals, contain large quantities of taurine. For example, about 1% of the dry weight of a rat is taurine. Most scientists reading this carry around with them in excess of 60 g of taurine.

The Function of Taurine in the Brain

What is the function of taurine? Questions concerning function are deceptively simple and infiltrated with linguistic and philosophic traps. " 'When I use a word,' Humpty Dumpty said in a rather scornful tone, 'It means just what I chose it to mean—neither more nor less.' " For a proper discussion of taurine in the brain, we must consider what it is we mean by such words as function or neurotransmitter. The urge to find out how things function is reinforced by the separation in our language between the doer (noun) and the deed (verb). A Japanese friend sent me a gift of an elegant wooden mascot doll called a "Kokeshi." I was amused to observe that the inevitable first question of anyone examining it was "What does it do?" Well, it did not do anything. It just was. In the same way, the verbal distinction between a substance and its "function" is artificial, representing an intellectual duality more than a dualism in nature in that a substance may only express its inherent reactivity in a given chemical environment. The substance and its properties are one and the same, in that we define and identify a substance by its properties. Thus, substances have been identified by melting point, spectrometric characteristics, solubility, ninhydrin reactivity, ability to contract guinea pig ileum, or to slow the beat of an isolated heart, to give examples ranging from physical to pharmacological properties.

"Function" is too useful a word not to use, even though some meanings of the word imply intent or purpose. These conceptions of the word are teleological, and thus have no place in a theory of evolution that does not accept the doctrine of final causes. In this sense, function implies a design to serve a given end, rather than a random progression of changes of which ones that tend to conserve the organism are perpetuated.

Another problem with the word is the tendency we have to define substances in terms of a single action or function. Testosterone is a sex hormone. Norepinephrine is a neurotransmitter. As was pointed out above, however, compounds often existed in the phylogenetic tree before the "function" evolved. As the organization of cells became more complex, more pathways of interaction and reaction became available for a given substance. Norepinephrine occurs in high concentrations in a number of cacti. No one would argue, however, that the giant cactus *Cereus giganteus* will eventually evolve a central nervous system. Taurine occurs in primitive organisms far below the phylogenetic point at which livers and bile salts arose. Unless we postulate a modern version of the doctrine of signatures, it is clear that the same molecule can not be subserving the same biochemical or physiological processes in all species in which it is found. Thus, statements of the form "norepinephrine is a neurotransmitter" are oversimplications, forced on us by the structure of the English language, in which every deed must have a doer, and we must cast our thoughts, and hence our perceptions of the world, into the mold of subject, verb, and object. As a result, we are constrained into statements such as "the lightning flashes," when a moment's thought reveals that the lightning and the flash are one and the same thing.

Returning to the original question "What is the function of taurine?", I have argued that in one sense the question, being teleological, is meaningless. In another sense, the differentiation between taurine and its function is invalid. No compound can be more or less functional than another. Whatever the environment, be it test tube or cell, a molecule can only express its inherent chemical reactivity in its interactions. Chemical laws are universal, however complex the environment. Biochemical processes are an inevitable result of the reactivities of the chemicals involved. A living cell can not confer on a molecule a property it does not also have outside the cell. To dissect out one particular molecular interaction and to call this process a function is again a duality of the mind but not of nature.

Molecules interact in such a manner as to lower the free energy of the system by the available reaction pathways. Some of the interactions in which a given substance may be involved in the cell may be necessary for the cell to endure. Without them, metabolism could not be sustained. Others may be needed for cell replication, and still others may permit the cell to respond to stress conditions, or to modulate such responses. There is no more reason these processes should be described by a word such as "function"—except for convenience—than that the reaction that occurs when sodium hydroxide is added to hydrochloric acid should be described as the function of sodium hydroxide. It is merely a purposeless expression of inherent chemical reactivity in the pertaining environment. If no reaction occurs when a substance is added to hydrochloric acid, then that substance cannot be sodium hydroxide. A substance is nothing more than the summation of its properties, and its functions can only be an inevitable result of its properties.

So what is meant by the question, "What is the function of taurine in the brain?" All phenomena at all levels—behavioral, physiological, or biochemical—cal—result from molecules interacting with molecules. What macroscopic phenomena result from the molecular interactions of taurine in the brain? How may the phenomenological consequences of taurine in the brain be investigated? Here we run into problems. The pharmacology of taurine is complex, and many things happen when taurine is introduced into the brain. It is difficult to argue backwards from such phenomena in order to deduce physiological processes. An increased level of taurine may exaggerate a physiological action or it may create new phenomena. Insight into the actions of taurine in the brain will follow from the availability of methods to lower levels of taurine or the development of compounds that will antagonize its action, as a comparison with GABA shows. Except for a few cases, these are not available.

Relationship Between Taurine and GABA

Taurine and GABA are similar substances both biochemically and chemically. The position of GABA in CNS phenomena is, however, more sharply defined. Considering that the two compounds are so similar, why is so much more known about the one than the other? Both substances are αω-amino acids, present in the CNS in high concentrations, and both have inhibitory actions on neuronal processes. The compounds differ in that GABA is metabolically and chemically unstable, whereas taurine is essentially nonmetabolized and is chemically stable. Furthermore, whereas GABA is limited in distribution to the CNS, taurine is present both in the CNS and in peripheral tissues. Our difference in insight into the physiology and biochemistry of these compounds is probably due to our lack of knowledge of the pharmacology of taurine, whereas the pharmacology of GABA has been studied intensively for several thousand years. The Godplant, soma, the pharmacology of which was described in detail in the Aryan vedas 3,500 years ago, has now been identified as the mushroom *Amanita muscaria,* and one active ingredient shown to be the GABA agonist muscimole [Wasson, 1979]. Many other agonists for GABA are now available although muscimole is still one of the most useful. The Nomadic tribes of Siberia, users of *A. muscaria* or fly agaric, know more about the pharmacology of GABA than we know of the pharmacology of taurine, in that we still do not have available a usable taurine agonist. Furthermore, numerous GABA antagonists, such as bicuculline, are known. A number of transport antagonists for taurine are known, but no pharmacological antagonists are yet available [Huxtable et al, 1979]. It is essential that workers in this area obtain both agonists and antagonists to taurine, because these are the fundamental tools needed for definitive studies on taurine in the CNS.

GABA is commonly accepted to be a neurotransmitter. Given the known similarities between the two compounds, is taurine also a neurotransmitter? Taurine is present both within and outside of the CNS. If it has a physiologically important role in both areas, and it is serving as a neurotransmitter in the CNS, then it is presumably serving some other function in peripheral tissues. This also suggests that in the central nervous system it may be polyfunctional, ie, serving as a neurotransmitter in addition to some other role. In terms of structure, taurine would be an excellent candidate for a neurotransmitter. It is a simple substance with a high dipole moment, and it is not utilized as a protein substrate. Most known neurotransmitters or neurotransmitter candidates are either amino acids or modified amino acids, such as serotonin, histamine, catecholamines, GABA, glycine, aspartate, glutamate, and the peptides. Even neuroactive compounds such as the opiates are nothing but modified dipeptides. If an inhibitory neurotransmitter merely carries a message instructing the postsynaptic membrane to hyperpolarize, no more elaborate a compound than taurine would be needed.

If the operational characteristics defining a neurotransmitter are examined, taurine fulfills these criteria as well as most other neurotransmitter candidates. It is present in neurons, there is a subpopulation of synaptosomes that binds and transports taurine with high affinity, there is a steep enough concentration gradient across the normally impermeable neuronal membrane that an increase in permeability would result in the release of large amounts of taurine, and it depresses neuronal firing. Taurine has a large number of inhibitory actions in the central nervous system.

Does all this mean that we should label taurine a neurotransmitter? My own view is no, because I am not sure we know what we mean when we talk of neurotransmission. The neurotransmitter theory is a hypothesis that has been very useful in accounting for a large number of observations, and in promoting numerous investigations in the CNS. However, it may have outlived its usefulness, and it may be too simple an explanation to account for the diversity of recorded phenomena. As has been recently pointed out, there are too many neurotransmitters [Anderson, 1980]. Furthermore, there are many more "putative" neurotransmitters than there are transmitters. This redundancy of potential, possible, or actual neurotransmitters suggests that the phenomena in which they are involved are more complex than just the simple carrying of a yes/no instruction. Possibly, many of these substances, particularly the most complex peptides, carry additional information than that of the transmission of an electrical state across a synaptic cleft. One scientist knowledgeable in the area suggests that there may be in excess of 200 peptide neurotransmitters [Snyder, 1980]. If this is the case, it is time to rethink the whole theory of neurotransmission.

Taurine Levels in the Brain

There are numerous measurements scattered through the literature of taurine concentrations in various areas of mammalian brain. Some representative values for brains of adults are shown in Table II. These figures exemplify two general observations: that taurine concentrations tend to be high, and that in most species there is a fairly marked regional variation in content. Access to the extensive literature on taurine distribution in the brain can be obtained through the list of major reviews given at the end.

Table III, which like Table II is a fairly random collection of data, gives an indication of how taurine concentrations compare to those of other neuroactive amino acids. In some cases, the concentration of taurine is exceeded only by that of glutamate. The function of taurine has been most intensively studied in

TABLE II. Taurine Levels in Adult Brains (μmol/g wet weight)

Brain Region	Mouse[a]	Rat[b]	Human[c]	Cat[d]	Guinea pig[e]
Cerebellum	9.1	4.7	3.3	2.7	1.1
Medulla	5.3				
Hypothalamus	4.8	2.5	1.1	1.8	1.0
Striatum	11.9				
Cerebral cortex	10.0	1.5	1.4	2.1	
Cerebral hemispheres		5.6			1.0
Midbrain		2.6			0.8
Pons medulla		1.7			0.9
Inferior colliculus		1.2			1.1

[a]HaM ICR Swiss mouse [Hruska et al, 1975].
[b]Seizure-susceptible Sprague–Dawley rat [Laird and Huxtable, 1976].
[c]Perry et al, 1971
[d]Guidotti et al, 1972.
[e]Huxtable and Lippincott [1980].

the retina, and although this area is beyond the scope of the present article, the high levels of taurine in retina compared to those of other amino acids are so astonishing as to be worth illustrating (Table III).

As Lombardini [1976] has pointed out, it is often difficult to compare data from different laboratories, due to differences in dissection techniques and analytical procedures. However, the figures on Tables II and III give a general indication of the varability of taurine from brain-to-brain, region-to-region, and relative to other amino acids. The comparison of taurine and GABA levels in adult brains is somewhat misleading, because of the opposed developmental characteristics of these substances. GABA levels tend to be low in fetal and neonatal brain tissue, and concentrations increase steadily through brain maturation (eg, see data quoted in [Himwich and Agrawal, 1969]). Taurine concentrations, however, tend to be extremely high in fetal brain tissues [Chanda and Himwich, 1970; Davis and Himwich, 1973]. Concentrations drop rapidly during the postnatal period, and adult concentrations may be only one-quarter to one-third those of the young animal [Agrawal et al, Kaczmarek, 1971]. This singularity has fueled much speculation concerning the possible involvement of taurine in brain maturation.

TABLE III. Relative Concentrations of Neuroactive Amino Acids in Brain and Retina (μmol/g wet weight)

	Rat[a]			Rat[b]		Human[c]	Neonatal brain[d]				Retina[e]		
	Frontal cortex	Cerebellum	Pons-medulla	Cortex	Inferior colliculus	Cerebellar hemispheres	Mouse	Guinea pig	Cat	Rabbit	Rat	Cat	Rabbit
TAU	5.6	4.9	2.0	5.5	2.2	2.2	16.2	3.4	9.2	5.9	50	43	52
ASP	2.9	3.1	3.4	3.1	3.3	1.6	1.9	2.8	1.1	1.2	3.1	2.9	2.2
GLU NH_2	2.1	3.3	1.7	2.8	2.6	5.4	3.8	4.9	3.9	3.5	5.2	3.6	4.8
GLU	10.1	10.9	5.6	10.3	6.1	10.1	4.3	11.0	6.0	5.2	10.5	6.4	13.2
GLY	1.4	1.5	3.5	0.7	1.6	2.0	2.7	2.6	2.3	3.2	2.2	2.0	2.1
ALA	1.2	1.5	1.3	0.6	0.4	2.1	3.3	0.9	2.2	0.9			
GABA	4.4	3.3	2.8	1.6	3.2	1.4	1.4	1.9	1.3	1.2	2.8	2.0	2.7

[a]Sprague–Dawley [Huxtable et al 1979].
[b]Seizure-susceptible rat [Huxtable and Laird, 1978].
[c]Perry and Hansen [1971].
[d]Himwich and Agrawal [1969].
[e]Voaden et al [1977].

Taurine Balance in the Brain

The levels of taurine attained in the brain are the sum of processes giving rise to taurine and processes removing it. These processes are the biosynthesis of taurine, the further metabolism of it, and the net movement of taurine into or out of the brain by transport. The current state of knowledge concerning these processes is reviewed in order to expose the glaring defects in our understanding. The more basic information that can be obtained on how taurine is handled by the brain and how its levels are regulated, the more likely we are to be able to deduce fundamental insights into its function and behavior.

In particular, we need information on the quantitative aspects. How much taurine is made by a given enzymatic pathway? How many micromoles a day are transported into and out of the brain? How do the rates of these processes vary during development and in disease or in response to nutritional stress or pharmacological agents? The biochemistry of taurine is a quagmire of speculation. A few islands of fact obtrude, but the relationship of each to the other is unproven or uncertain at best. Individual enzymes have received systematic study by one or more groups but almost no effort has been directed toward the quantitative aspects of these enzymes in the flow of sulfur compounds through the brain.

Biosynthesis

Direct evidence that the brain makes taurine is sparse. Rat brain slices or homogenates incubated with ^{35}S-methionine have been shown by paper chromatography to produce radiolabeled cysteine, cysteine sulfinic acid (CSA), cysteic acid (CA), hypotaurine, and taurine [Peck and Awapara, 1967]. Collins [1974] further showed in rats that ^{14}C-cysteine given ICV resulted in ^{14}C-taurine in various areas of the brain. The percentage of activity in the various structures that was present as taurine ranged from 7% in the hypothalamus to 19% in the olfactory bulb. A reasonably good correlation was obtained between these percentage conversions and the indigenous taurine contents of these regions. No intermediates were identified. A further refinement of this experiment was the simultaneous administration of ^{3}H-cysteine by the intraperitoneal route and ^{35}S-cysteine by the ICV route in rats. The incorporation of ^{35}S into taurine was observed within 1 hr in the brain, and incorporation reached a maximum by 4 hr. Incorporation of ^{3}H into taurine in the brain, however, was not a maximum until 3 days following injection. This is a sufficient demonstration that the conversion of cysteine to taurine when given ICV is, in fact, occurring in the brain, and not in some peripheral organ followed by reexport to the brain [Pasantes-Morales et al, 1980].

These studies show the brain has the capability of making taurine. Major problems with these investigations, however, are that only the rat was examined, and none of the experimental approaches yield quantitative information as to the rates of taurine synthesis.

There are a number of potential pathways by which cysteine may be converted to taurine in the brain. One such pathway involves the conversion of cysteine to cysteamine, and the further conversion of cysteamine to hypotaurine and taurine (Fig. 3). The conversion of cysteine to cysteamine has been shown to occur in a number of tissues. However, the enzymology has yet to be worked out. Although cysteamine is merely the decarboxylated analog of cysteine, no decarboxylase performing a direct reaction has been found, despite searches by a number of groups. The metabolism of cysteine to cysteamine appears to involve a number of complex steps including the intermediacy of posphopantothenoyl cysteine [Huxtable, 1978]. Supernatants of brain homogenates of rat, rabbit, and chicken have been shown to have a finite but relatively low ability to produce $[^{14}C]CO_2$ from $[U-^{14}C$-cysteine]phosphopantothenoyl cysteine. The formation of radioactive hypotaurine and taurine was also shown. Formation rates ranged from 450 nmol/hr/g of protein in the rabbit to 250 in the chicken. Dog brain was without activity [Scandurra et al, 1977].

Cysteamine dioxygenase, the enzyme catalyzing the conversion of cysteamine to hypotaurine, has been well characterized in a number of tissues, including liver and kidney. The enzyme has been purified, and its biochemical parameters have been determined [Cavallini et al, 1971]. However, I cannot find any evidence for the direct measurement of cysteamine dioxygenase activity in brain.

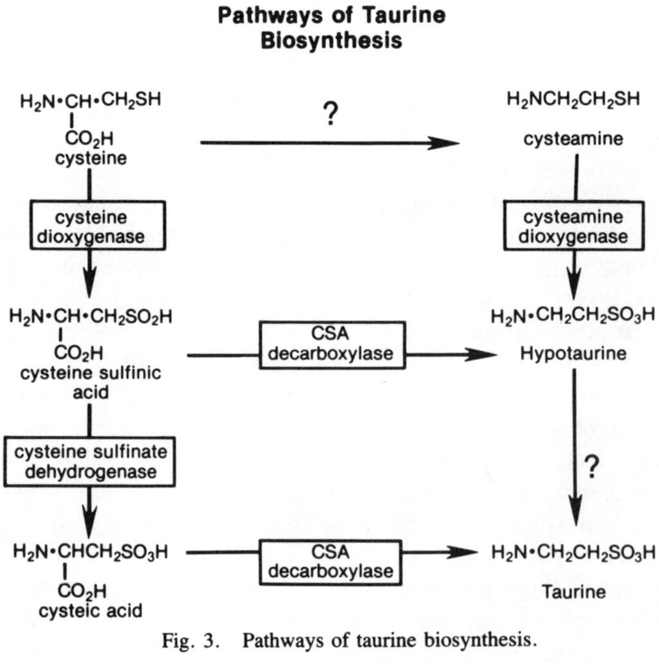

Fig. 3. Pathways of taurine biosynthesis.

The last step of the sequence, the conversion of hypotaurine to taurine, also presents major biochemical problems, as is discussed further below.

Another proposed pathway of biosynthesis in the brain involves the direct incorporation of inorganic sulfate, via the intermediacy of phosphoadenosine phosphosulfate (PAPS), into CA and taurine. The evidence that this pathway exists is tenuous to say the least, and in considering it I am reminded of a complaint made over 200 years ago by a Shakespearian editor faced with the efforts of previous editors; "It is no pleasure to me. . . . to observe how much paper is wasted in confutation. Whoever considers the revolution of learning. . . . must lament the unsuccessfulness of inquiry, and the slow advances of truth, when he reflects that the great part of the labor of every writer is only the destruction of those that went before him." Conversion of ^{35}S-PAPS to ^{35}S-taurine has been claimed to occur in the liver and heart of numerous mammals [Martin et al, 1974]. However, one of the proposed intermediates, amino acrylic acid [Sass and Martin, 1972], is unstable in water. The addition of nonlabeled CA to incubates not only does not alter the incorporation of ^{35}S into taurine, as would be expected if CA was an intermediate in this step, but it stimulates incorporation by some 40% [Hill and Martin, 1973], and measured incorporation of radiolabel from ^{35}S-PAPS is markedly depressed by cysteine, S-adenosyl methionine, CSA, and taurine in the chicken liver when these substances are present at physiological concentrations [Miraglia et al, 1966; Hill and Martin, 1973]. Furthermore, the rates of supposed conversion of sulfate to taurine are so low as to be of little physiological significance. For example, 6 hr after rats were given 20 μCi [^{35}S]SO_4 by stomach tube, the apparent incorporation into taurine in the liver was 147 cpg of liver (sic) [Martin et al, 1972]. Others have not been able to show any incorporation of [^{35}S]SO_4 into taurine or other organic molecules in in vitro incubations of mammalian tissues [Dziewiatkowski, 1954; Fromageot and Chapeville, 1956; Pearson, 1956; Green and Robinson, 1960]. The small incorporations occasionally detected in in vivo experiments have been ascribed to gut bacterial metabolism, as germ-free rats show no incorporation [Huovinen and Gustafeson, 1966; quoted in Jacobsen and Smith, 1968]. The cat is supposed to have as high a level of PAPS transferase as the rat [Martin et al, 1974], yet intravenous administration of 3 mCi [^{35}S]SO_4 to each of three cats did not result in any detectable radioactivity in taurine 5 days later [Knopf et al, 1977].

Although the evidence for the existence of the PAPS pathway to taurine in mammals has been heavily criticized a number of times [Huxtable, 1976; Knopf et al, 1977], this pathway is still being quoted as one of the established routes to taurine [Oja et al, 1977]. The criticisms of it are explicit. The burden of proof lies with the proposers of this pathway as a significant source of taurine in mammals.

The biosynthetic pathway to which most attention has been paid involves the sequence cysteine, CSA, hypotaurine, and taurine (Fig. 3). The first intermediate in this step, CSA, has been shown to be present in rat brain [Bergeret and Chatagner, 1954]. The enzyme catalyzing its formation is cysteine dioxygenase, and this has been found both in liver and brain [Yamaguchi et al, 1973; Yamaguchi, 1980]. Peck and Awapara [1967] were the first to show the conversion of ^{35}S-methionine to CSA in the brain. Early workers were unable to find enzymatic activity [Wainer, 1964, 1965; Ewetz and Sörbo, 1966]. However, subsequent workers have found high activity in the central nervous system [Yamaguchi et al, 1973; Misra and Olney, 1975]. There is a regional distribution of activity, the highest levels being found in the hypothalamus and superior colliculi, and the lowest in the hippocampus [Pasantes-Morales et al, 1977]. The enzyme is NAD^+ dependent and is localized in microsomes and synaptosomes. There are interesting developmental changes in enzyme activity. Levels are low in rats at birth, but increase fivefold during the postnatal development. There is also a threefold difference in activity in different brain regions, with pons medulla being highest and cerebral cortex lowest [Misra and Olney, 1975]. Similar developmental changes have been observed in rat retina [DiGiorgio et al, 1975].

The next step in this pathway involves the decarboxylation of CSA to hypotaurine. Hypotaurine occurs in the brain in rather small amounts [Perry and Hansen, 1973]. The enzyme catalyzing the conversion, CSA decarboxylase, has been shown to be present in high amounts in rat, dog, and cat brain, and in lower amounts in man and rabbit brain [Jacobsen and Smith, 1963; Jacobsen et al, 1964]. In adult rats, it was originally claimed that enzyme activity was rather evenly distributed throughout the central nervous system [Kaczmarek et al, 1971], but a more recent study of activity in 12 regions of the rat brain revealed a sixfold difference in levels, with the superior colliculus being highest, and the pons medulla being lowest [Pasantes-Morales et al, 1977]. Most of the activity was located in the synaptosomal fraction. Activity was determined by measuring the release of $[^{14}C]CO_2$ from labeled CSA. This is not such a reliable method as is direct demonstration of the products of enzymatic reaction, because other metabolic processes may be intervening. For example, homogenates of mammalian brain release $[^{14}C]CO_2$ from ^{14}C-aspartic acid, even though no aspartate-4-carboxylyase is present [Gaull and Rassin, 1980].

There is an interesting developmental change in CSA decarboxylase activity in rats, in that activity at birth is very low, but increases tenfold postnatally [Agrawal et al, 1971; Oja et al, 1973]. The same postnatal increase is observed in cat brain, in which there is also heterogenous distribution of activity [Piha and Saukkonnen, 1966; quoted in Oja et al, 1977]. A peculiarity is the lack of correlation between the regional distribution of taurine and CSA decarboxylase in brain [Agrawal et al, 1971]. Furthermore, the levels of taurine decrease rather sharply in the brain during development.

The enzyme, like many decarboxylases, is pyridoxine dependent [Rassin and Sturman, 1975]. Brain homogenates prepared from pyridoxine-deficient rats show no activity unless pyridoxine is added to the incubation. The enzyme is somewhat nonspecific, also decarboxylating CA. Glutamic acid is a noncompetitive inhibitor [Rassin and Sturman, 1975].

The final step in this pathway is the oxidation of hypotaurine to taurine by a so-called "hypotaurine dehydrogenase." This is a mysterious step in that no one has been able to characterize the enzyme. It was named, and given an enzyme control number, on the basis of a poor study carried out on liver homogenates [Sumizu, 1962]. The following comment has been made about this report: "It is unfortunate that such a badly conducted and controlled study still misleads many investigators into believing the verified oxidation of hypotaurine" [Oja et al, 1977]. This work has not been confirmed by other investigators (eg [Fiori and Costa, 1969]). Hypotaurine given in vivo is converted to taurine [Huxtable, 1976; Kontro and Oja, 1980], and hypotaurine is present in the brain, albeit at concentrations only 1% of those of taurine [Perry and Hansen, 1973]. Past this point, there are no convincing studies on enzymatic conversion in the brain. One group [Kontro, 1980; Kontro and Oja, 1980] claims to have shown the enzymatic conversion of hypotaurine to taurine in homogenates of mouse brain. However, the rates they have observed are so low as to be unconvincing. Incubation of homogenates with 50 μM ^{35}S-hypotaurine results in less than 1% conversion to taurine [Kontro, 1980]. If the rate of conversion is this low, why are taurine levels so high in the brain, whereas hypotaurine levels are so low? In whole rat brain, Perry and Hansen [1973] found taurine levels of 6.6 μmol/g and hypotaurine levels of 0.07. As CSA decarboxylase activities are so high, one would expect an accumulation of hypotaurine.

There is a further potential biosynthetic route to taurine in the brain which deserves more consideration than it has received. This is a route involving the conversion of CSA to CA and to taurine. Peck and Awapara [1967] have shown that ^{35}S-methionine can give rise to ^{35}S-CA in brain homogenates. However, the enzymology of conversion of CSA to CA is unclear, and has not been well studied. When subcellular fractions containing cysteine dioxygenase were incubated with radiolabeled cysteine, both CSA and CA were produced in an approximately 3:1 ratio [Misra and Olney, 1975]. This might indicate that the preparation also contained cysteine sulfinate dehydrogenase activity.

The conversion of CA to taurine has been better studied. As mentioned above, most workers have concluded that CSA decarboxylase will use both CSA and CA as substrate, the Km for the latter being 3 to 10 times higher [Jacobsen and Smith, 1968]. However, there are some intriguing studies on liver which suggest that there perhaps may be a separate enzyme catalyzing decarboxylation of CA. In general, it is difficult to separate decarboxylases because of their similar physical chemical properties. An early study on dog liver found that CSA decarboxylase activity purified away from CA decarboxylase activity [Sorbo and

Heyman, 1957]. Later workers purified a CSA decarboxylase from liver which had no activity for CA at all. Whole homogenate decarboxylated CSA 2.4 times faster than CA. The purified enzyme was 180 times more active in decarboxylating CSA, but did not react with CA [Lin et al, 1971].

The intact rat rapidly converts exogenous CA to taurine. When rats are given a diet containing 0.1% CA, plasma taurine levels increase sixfold [Anderson et al, 1976].

In the brain, however, the distribution of the two enzymatic activities appears to be the same [Pasantes-Morales et al, 1977]. In mouse brain, there is a marked regional distribution of CA decarboxylase activity, with the pons medulla having only 10% of the activity found in the putamen. In this study, ^{35}S-CA was used as substrate, and the production of ^{35}S-taurine was measured directly. This is an improvement over measuring release of carbon dioxide as an index of decarboxylase activity [Wu et al, 1979]. The same group found two forms of CA decarboxylase in bovine brain. One form decarboxylated CSA and CA with the activity ratio remaining the same on purification. Furthermore, each substrate competitively inhibited the decarboxylation of the other. The other form, CA decarboxylase II, also decarboxylated glutamic acid. Further work is needed to determine whether or not this enzyme is identical to glutamic acid decarboxylase [Wu et al, 1978].

Most workers accept the CSA decarboxylase route as being the most important in taurine biosynthesis in the brain. However, there are major objections to this route, some of which may be removed by further studies. There is no correlation in distribution between CSA decarboxylase and taurine. Enzyme activities increase with age, whereas taurine levels decrease. Despite fairly intensive studies, there is no convincing evidence that an enzyme exists for oxidizing hypotaurine to taurine. It could be that the more important route is the one through CA. The fact that the Km for CSA decarboxylase is lower than for CA decarboxylase is not necessarily of much import. Only one metabolic reaction has been found for CA, decarboxylation to taurine. However, CSA can be metabolized in several ways, in keeping with the sulfur atom being present at an unstable oxidation level (Fig. 4). It may be oxidized to CA, decarboxylated to hypotaurine, or transaminated by CSA transaminase to β-sulfinyl pyruvate and hence to pyruvic acid, while at the same time converting α-keto-glutarate to glutamic acid. In fact, CSA transaminase activity is much higher than CSA decarboxylase activity in olfactory bulb of rat [Gabellec et al, 1978; Austin et al, 1978]. Transamination of CSA with α-keto acids also occurs very readily in rat liver mitochrondria. These observations are consistent with oxidation of cysteine to CSA and conversion of CSA to inorganic sulfate via CSA transaminase being the major catabolic route for sulfur amino acids. Thus, how much CSA is metabolized to hypotaurine is a function not only of the Km of this process but of the competing Kms of CSA dehydrogenase and CSA transaminase. More information is needed on the levels and distribution of CA in the brain, and on CSA dehydrogenase. However, in the light of the serious problems in establishing the CSA decarboxylase pathway as a major source of taurine, the CA route must be considered a viable alternative worthy of much more study. Establishment of this route as quantitatively important would cut the Gordian knot presented by hypotaurine which so many of us have been fruitlessly endeavoring to unravel.

The possibility that biosynthesis of taurine in the brain may be of little quantitative significance under normal conditions must not be overlooked. The relative significance of biosynthesis and diet as sources of taurine in the intact animal is discussed below, and some of the data are intriguing. In the cerebellum of the adult rat, for example, about 70% of the taurine is derived from the diet when this contains about 5 μmol/g of taurine. Only 30% is derived from biosynthesis, but the site of biosynthesis is unknown. It may be in the cerebellum or in the liver followed by export to the cerebellum or a combination of the two.

CSA may also be intimately involved in the regulation of brain excitability. Acidic amino acids, such as glutamate, aspartate, CSA, and CA are neuroexcitatory, whereas neutral αω amino acids, such as hypotaurine, taurine, and GABA are neuroinhibitory. Of the three enzymes that accept CSA as substrate, one (CSA transaminase) produces another excitatory amino acid (glutamate), one produces an excitatory acid (CA) that is metabolized further to an inhibitory acid (taurine), and the third enzyme, CSA decarboxylase, yields an inhibitory

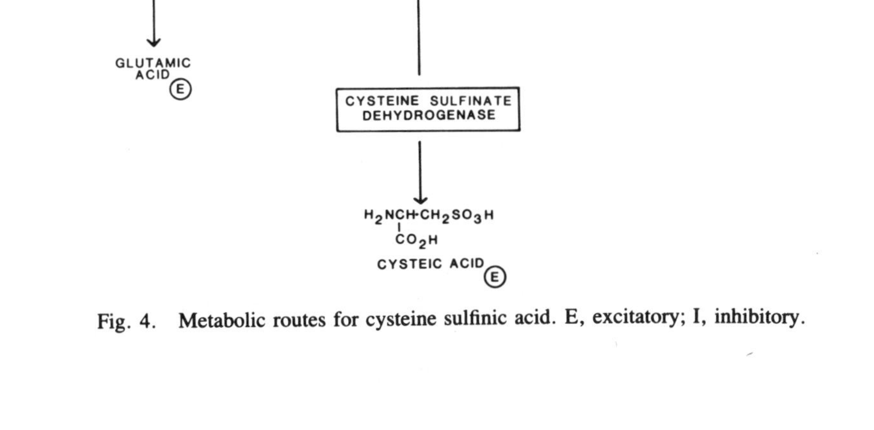

Fig. 4. Metabolic routes for cysteine sulfinic acid. E, excitatory; I, inhibitory.

amino acid. It may be that in the liver these enzymes serve primarily a metabolic function, whereas in the brain they have a primary role in the regulation of excitability.

Metabolism of Taurine

The proposed metabolites of taurine in mammals include taurocholic acid, isethionic acid (ISA), guanidinoethyl sulfonate (taurocyamine), and inorganic sulfate. The first of these, taurocholic acid, is not formed and does not occur in the brain. In order to keep the review within a reasonable length, I will not discuss the involvement of sulfate and guanidino compounds in taurine metabolism. However, because the literature on ISA is contentious, and some theories on the physiological actions of taurine have involved its metabolism to this substance, I will briefly review this area. Research on ISA revolves around two questions: Does ISA occur in mammalian tissues? Is it formed from taurine?

The evidence for the occurrence of ISA in mammals is convincing, but there is disagreement as to the levels. Jacobsen et al [1967] have measured the urinary excretion of ISA from man as being 108 μmole/24 hr. ISA was also determined to be in dog heart at a concentration of 1 μmol/g, rat heart at 3.4 μmol/g, and rat cerebral cortex at 3 μmol/g [Welty and Read, 1962; Hoskin and Brandes, 1973]. These and other reports aroused considerable interest, because they appeared to offer a key to the understanding of taurine in excitable tissues. In marine arthropods, ISA is the major axonal anion, with concentrations approaching 220 μmole/ml in the squid giant axon [Koechlin, 1955]. Furthermore, if it were a metabolic product of taurine, in forming it a zwitterionic substance would be converted to a strong anion. Several variations of a theory were proposed to account for the action of taurine on membrane excitability and ion flux in terms of the need to neutralize ISA formed within the cell with a cation such as potassium or calcium [Read and Welty, 1962; Chazov et al, 1974; See discussion in Huxtable, 1976]. However, the methods of assay used in these papers were open to criticism, and recent work has not substantiated these high levels. In one study, ISA was isolated by ion exchange chromatography on cationic AG50 resin, in which it eluted later than taurine [Welty et al, 1962]. This is in defiance of chemical common sense and contradicts the findings of others on the chromatographic behavior of ISA, in which this substance elutes essentially in the void volume of such a column [Peck and Awapara, 1967; Huxtable and Bressler, 1972; Urquhart et al, 1974]. Another assay method involved the turbidometric determination of barium sulfate formed by oxidation of ISA [Hoskin and Brandes, 1973]. The presence of these high levels of ISA have not been substantiated by more sophisticated investigations. Analytical methods such as GC-MS have shown the levels in rat heart to range between 0.008 and 0.2 μmole/g, in dog heart to be about 0.3 μmole/g, and in rat brain to be 0.016 μmole/g [Fellman et al, 1978; Schaffer et al, 1978; Remtulla et al, 1977].

Is ISA formed from taurine? Brain slices were reported to convert ^{35}S-taurine to ^{35}S-ISA [Peck and Awapara, 1967], and intraperitoneal injections of tracer amounts of ^{14}C-taurine to the intact rat was shown to give rise to small quantities of ISA in all tissues examined. Five days following injection, for example, 2% of the total radioactivity in the brain was in the form of ISA [Huxtable and Bressler, 1972]. Urquhart et al [1974] found a radioactive substance in the brain eluting off an amino acid analyzer in the same position as ISA following parenteral administration of ^{35}S-taurine to rats and monkeys. However, the kinetics of elimination of ISA from the brain are quite different from the kinetics of elimination of taurine, suggesting that in quantitative terms most of the taurine leaving the brain is not converted to ISA first [Lombardini and Homan, 1979]. The conversion of taurine via sulfoacetaldehyde to ISA has been well established in bacteria, and Frederici and co-workers [1974] examined whether the small conversions occurring in rats were due to the intestinal flora. They injected ^{35}S-taurine to two groups of rats, one group of which was receiving sulfaguanidine. The antibiotic did not affect the excretion of radioactivity. The appearance of radioactivity in ISA, as a percentage of total radioactivity excreted that day, varied from 5% the first day to 18% on the third day. Sulfate excretion varied from 6% on Day 1 to 47% on Day 2, and taurine from 89% on Day 1 to 42% on Day 2. Where the conversion occurs is still unresolved. Fellman et al [1978] have been unable to demonstrate conversion of taurine to ISA in homogenates and slices of heart, brain, and liver.

Further investigations on ISA are needed. However, the current state of knowledge can be summarized as follows. ISA occurs in rather small amounts in brain. It may or may not be formed from taurine, but if so this is not a quantitatively important route for the metabolism of taurine. Other routes by which ISA may be formed not involving the intermediacy of taurine have been proposed, and partially studied [Scandurra et al, 1978; Frederici et al, 1974]. Finally, the physiological actions of taurine are not predicated on its conversion to ISA.

Transport of Taurine

It has been abundantly demonstrated that brain slices can take up taurine, although in none of these studies were data obtained as to the quantity [Oja, 1971; Kaczmarek and Davison, 1972; Lahdesmaki and Oja, 1972; Honegger et al, 1973; Collins, 1974; Benjamin and Quastel, 1977 and others]. It appears that both neuronal and glial cells can transport taurine [Borg et al, 1976; Borg et al, 1980; Balcar et al, 1980]. Brain astrocytes, used as models for normal glial cells, exhibited a V_{max} of transport much higher than that found for synaptosomes [Schousboe et al, 1976], suggesting that glia may be quantitatively important for the removal of taurine released at synaptic terminals.

It has also been abundantly demonstrated that brain synaptosomes have a high affinity transport system for taurine. This area has been reviewed by Lombardini [1978]. Hruska et al [1978b], in a thorough recent study, have fully characterized the taurine transport system in rat synaptosomes. They found it to be a dual affinity system, exhibiting a high-affinity Km of 3.2 μM and a low-affinity Km of 3340 μM. These are corrected Kms, and the importance of these corrections have been stressed in a recent review [Hruska et al, 1978a].

One of the most interesting aspects of taurine transport is that in tissues containing β1-adrenergic receptor sites, such as heart, salivary gland and pineal, taurine movements are accelerated by β-adrenergic agonists or other agents that increase cellular cyclic AMP concentrations [Huxtable and Chubb, 1977; Chubb and Huxtable, 1978; Huxtable et al, Azari, 1980; Azari and Huxtable, 1980; Wheler et al, 1979; Wheler and Klein, 1980].

I suspect that this relationship is a fundamental part of the physiological action of taurine, and that the area of the interaction between taurine and the adrenergic amines will prove very fruitful over the next few years.

Taurine Economy of the Brain

Quantitation of the source of taurine has been become an issue of medical as well as of scientific interest. Alterations in taurine level occur in congestive heart failure in rats, dogs, and humans [Peterson et al, 1973; Huxtable and Bressler, 1974a,b]. Taurine stress occurs in animals or patients receiving cholestyramine, in which resorption of bile salts is decreased [Cook et al, 1971; Howard and Courtenay Evans, 1974]. Cats kept on a taurine-deficient diet develop retinopathy leading to blindness [Schmidt et al, 1976; Knopf et al, 1978]. Questions as to the essentially of taurine in the human diet have recently been raised [Gaull et al, 1977]. This is an issue in pediatric nutrition, as the cow-based formulae milks that most babies in Western nations receive contain no taurine. In these infants, a large number of biochemical abnormalities may be detected, although no clinical problem has yet been found associated with this. This may be because babies are placed on taurine-containing food stuffs before deficiency states may be produced, or it may be that no one has examined closely enough for deficiency states. If taurine is an essential nutritional element for humans, this also raises questions concerning taurine deficiencies in adults. Although most diets in the industrialized nations contain excessive amounts of taurine, humans with kidney resorption defects, or people who are strict vegetarians, may develop taurine deficiency states. Defects in resorption of taurine have been observed in substrains of mice [Blake et al, 1974; Chesney et al, 1976, 1978], and in patients suffering from Friedreich's ataxia, myotonic dystrophy, and several other disorders [Lemieux et al, 1976; Filla et al, 1974]. Furthermore, some humans may require larger than average amounts of taurine. In several studies in various countries, numbers of patients with otherwise intractable epilepsies have been helped by taurine therapy (see Table VIII). In these cases, dietary intake of taurine was normal but increasing intake was sufficient to mitigate the epilepsy.

Many radiotracer experiments have been carried out with taurine in intact animals. Typically, these have involved giving injections of small quantities of radiolabeled taurine and following the distribution of label through the various organs of the body and its elimination from the body (eg [Awapara and Manz, 1957; Sturman, 1973; Collins, 1974; Lefauconnier et al, 1978]). This type of investigation has given valuable kinetic information. Data have been derived concerning the compartmentalization of taurine in the body, and rates of turnover from various organs. However, these experiments give no quantitative information as to the source of taurine. Our approach to this problem is illustrated in Table IV. Mice in our animal facility receive chow containing approximately 5 μmole/g of taurine. The amount of taurine in the brain that was derived from this source was estimated by placing mice on taurine-free diet, and giving them drinking water containing [^{3}H]taurine in a concentration such that their normal daily intake of taurine was maintained. From the radioactivity in the brain, the amount of taurine derived from the diet since the initiation of the experiment can readily be measured. As is shown in Table IV, equilibrium is attained in about 1 month, at which time, for most areas, approximately 40% of the taurine comes from the diet. The inferior colliculus receives a somewhat higher percentage, and also takes up dietary taurine at a faster rate.

The results of a similar experiment on the rat are shown in Table V. More of a variation between areas is observed, and the equilibrium state takes longer to achieve. These data are of great interest, inasmuch as the rat has a high

TABLE IV. Contribution of Dietary Taurine to Brain Taurine Content of the Mouse

	Percentage derived from diet[a]					
Days	2	4	7	10	21	30
Frontal cortex	7.6	11.1	17.3	21.1	31.8	39.2
Cerebellum	9.1	14.0	23.8	23.5	34.3	39.3
Cerebral hemispheres and midbrain	5.4	10.6	18.9	24.1	33.0	38.3
Inferior colliculus	29.7	40.4	37.2	61.6	46.5	57.1
Hypothalamus	13.1	17.6	26.0	26.4	39.8	37.8

[a]Mice were kept on a taurine-free diet and on drinking water containing 5 μM [^{3}H]taurine. Days are days they were kept under these conditions immediately prior to sacrifice. From the specific activity of taurine in the brain, the proportion derived from the diet was calculated. Two animals per point, from Huxtable [unpublished].

capacity for taurine biosynthesis. Despite this, when an exogenous source of taurine is available, the cerebellum obtains 70% or more of its taurine from this source. Thus, under normal dietary conditions, dietary taurine may be a major contributor to the taurine pools within the animal, despite a high biosynthetic capability.

A complementary experiment is illustrated in Table VI, in which rats have been maintained on a synthetic amino acid diet for the periods of time indicated in which ^{35}S-methionine was the only source of sulfur amino acids. The percentage of taurine that has been biosynthesized over the time course of the experiment is shown. The experiment is not the exact equivalent of that illustrated in Table V, in which animals were receiving a protein diet. However, the data are in good agreement. Of course, as this is an in vivo experiment, no indication is obtained as to where the taurine is biosynthesized. It may have been biosynthesized within the brain, or in a peripheral tissue and transported to the brain.

The taurine economy of neonates is possibly of even greater interest (eg, See [Sturman et al, 1977; Gaull and Rassin, 1979, 1980]. We determined how much taurine rats receive from the mothers in utero and during nursing by maintaining the dams on a diet containing [^{3}H]taurine before and during pregnancy, and during lactation. The accumulation of radioactivity in the brains of pups is shown in Figure 5, and the total taurine in the brain in Figure 6. From the difference in specific activity of the taurine in the pups and in the mothers, it can be calculated that at birth the brains contained 2.1 μmole of taurine derived from the mother, and at weaning 4.6 μmole. Corresponding figures for biosynthetically derived taurine are 0.1 and 8.9 μmole. Pups were weaned onto either taurine-containing or taurine-deficient diets, and the kinetics of elimination of radioactive taurine from the brain were followed in each case. It was a surprise to find that the half-life of exchange from the brain was not materially affected by the taurine content of the diet. Animals on a taurine-enriched diet had a half-life of 5.5 days, and those on the deficient diet had a half-life of exchange of 6.2 days. Taurine-deprived juvenile rats, therefore, do not respond by becoming avid conservers. No difference in brain concentrations of taurine could be found between rats on the deprived and enriched diet, indicating that the concentration is closely regulated, and that dietary intake, biosynthetic rate, and elimination mutually interact to achieve this regulation. These data are in disagreement with claims that in the neonatal rat, brain taurine turns over very slowly. In one study, the half-life of taurine in newborn rats was measured at 44 days [Sturman et al, 1977b]. Our finding that, during the period of nursing, biosynthesis of taurine is occurring is also in disagreement with others. It has been generally assumed that the major route of taurine biosynthesis in the rat is via the enzyme CSA decarboxylase. Kaczmarek [1976] has shown that little activity of CSA decar-

TABLE V. Contribution of Dietary Taurine to Brain Taurine Content of the Rat

	Percentage derived from diet[a]					
Days	4	7	10	15	30	60
Pons medulla	15.7	26.3	24.6	27.0	22.7	25.6
Midbrain	40.3	35.7	60.1	73.6	83.4	59.2
Cerebellum	17.1	28.4	38.2	38.8	62.1	69.9
Cerebral hemispheres	8.6	20.4	30.4	35.5	45.4	45.2

[a]Conditions as for Table IV. Three animals per point, from Laird et al [1980] and Huxtable [unpublished].

TABLE VI. Contribution of Biosynthesized Taurine to Brain Taurine Content of the Rat

	Taurine[a] (μmole/g wet weight)			
Days	4	7	11	15
Frontal cortex	0.88	1.25	1.92	2.37
Cerebellum	0.74	0.95	1.60	1.75
Cerebral hemispheres	0.62	0.91	1.58	1.78
Midbrain	0.66	1.01	1.69	2.07
Pons medulla	0.33	0.42	0.77	0.94
Inferior colliculus	0.20	0.56	1.03	1.44
Hypothalamus	0.58	0.55	0.93	1.60
Spinal cord	0.23	0.43	0.64	0.75

[a]Data are means of three rats per time point. Rats were maintained on a synthetic amino acid diet containing ^{35}S-methionine but no cysteine. Days are the days they were on this diet. From the specific radioactivity in taurine, the quantity derived from dietary methionine could be calculated, data from Huxtable [unpublished].

boxylase can be detected at birth, and activity is not appreciable until 30 days. Others have concluded that de novo synthesis in the rat is low up to 18 days following birth [Sturman et al, 1977c].

Of possible relevance to the human situation is the indication that much more taurine in the whole animal, and in the brain, is obtained from the mother via the milk than in utero. At birth, the rat has retained 27 μmole of maternal taurine, and by weaning this has increased to 120 μmole [Huxtable, 1980]. This is in agreement with the extremely high levels of taurine found in milk during the first few days of lactation [Sturman et al, 1977c; Rassin et al, 1978]. Human milk at the beginning of lactation contains in excess of 0.4 μmol/ml of taurine, or 21% of the total amino acids [Rassin et al, 1977; Rassin et al, 1978].

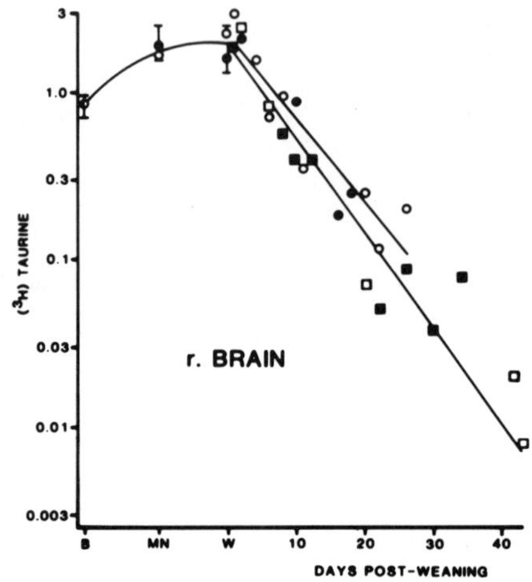

Fig. 5. Brain taurine in rat pups derived from the milk. The mothers had been equilibrated with [^{3}H]taurine before being bred and were maintained on a diet containing [^{3}H]taurine throughout pregnancy and lactation. Radioactivity in pup brains is shown arbitrarily as μmol[^{3}H]taurine/brain based on the specific activity the mother received. B, birth; MN, midnursing; W, weaning. Filled symbols are male animals, open females. Circles are animals weaned onto a taurine-free diet (half-life of taurine loss is 6.2 days, R = −0.94), and squares animals weaned onto a diet containing 0.4% taurine (half-life of taurine loss is 5.5 days, R = −0.94). Data from Huxtable [1980c].

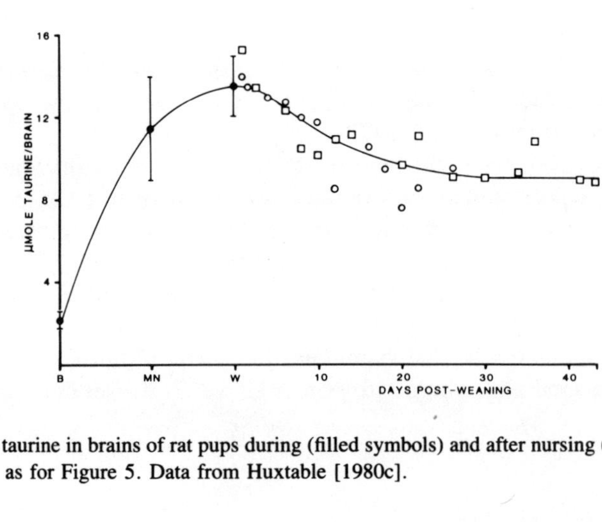

Fig. 6. Total taurine in brains of rat pups during (filled symbols) and after nursing (open symbols). Other symbols as for Figure 5. Data from Huxtable [1980c].

Effect of Guanidinoethyl Sulfonate on Taurine Concentrations in the Brain

It has been pointed out above that one of the experimental difficulties in the study of taurine in the brain has been the lack of methods for modifying its concentration [Huxtable, 1979, 1980a]. It has been shown that the amidino analog of taurine, guanidinoethyl sulfonate, is a competitive inhibitor of taurine transport in brain synaptosomes and the isolated perfused heart of the rat [Azari et al, 1979; Hruska et al, 1978; Huxtable et al, 1979]. Such a compound offered the possibility of lowering organ taurine concentrations in the intact animal by decreasing uptake from the circulation. Rats and mice maintained on 1% guanidinoethyl sulfonate in the drinking water show marked depletions in brain taurine concentrations over a period of a few days (Tables VII and VIII). Guinea pigs

TABLE VII. Effect of Guanidinoethyl Sulfonate on Brain Taurine Concentrations in Mouse*

	Percentage of control			
Days	5	9	15	20
Cerebellum	75.0	72.6	69.8	63.7
Cerebral hemisphere	66.0	69.8	61.3	49.6
Spinal cord	83.6	70.2	70.4	79.6

*Mice given drinking water containing 1% guanidinoethyl sulfonate for indicated number of days before sacrifice. Data from Huxtable and Lippincott [1980]; Huxtable et al, [1980]; and Huxtable [unpublished].

TABLE VIII. Effect of Guanidinoethyl Sulfonate on Brain Taurine Concentrations in Rat*

	Percentage of control		
Days	4	9	30
Cerebellum	78	70	43
Inferior colliculus	82	70	39
Hypothalamus	81	63	40
Frontal cortex	67	45	53
Pons medulla	74	57	26
Cerebral hemispheres	72	56	50
Midbrain	78	49	50
Spinal cord	67	84	47

*Rats given drinking water containing 1% guanidinoethyl sulfonate for indicated number of days before sacrifice. Data from Huxtable, et al [1979]; Huxtable et al [1980]; and Huxtable [1980b and unpublished].

and cats maintained similarly do not show changes in taurine concentrations [Huxtable and Lippincott, 1980]. These species, however, have the capability of metabolizing guanidinoethyl sulfonate to taurine. Guanidinoethyl sulfonate promises to be a valuable tool in the study of the functionality of taurine. It is to be hoped that better tools will be obtained, however, as guanidinoethyl sulfonate suffers from the disadvantages of not being effective in all species, and of being transported in place of taurine. A more suitable inhibitor would be one that blocked taurine transport without being transported itself. However, the action of guanidinoethyl sulfonate is remarkedly specific. Taurine concentrations may be reduced to 20% of control levels in certain tissues, without any effect being observed on the concentrations of other free amino acids [Huxtable et al, 1979].

The pharmacology of guanidinoethyl sulfonate is complex. Although it is a competitive inhibitor of taurine transport and mimics taurine in being transported by the system that is specific for β-amino acids, its pharmacological actions in the CNS are those of a taurine antagonist. In a number of species, it is epileptogenic. Furthermore, whereas intracerebral injections of 9.2 μmole of taurine into seizure-susceptible rats causes marked attenuation of seizure severity, similar injections of guanidinoethyl sulfonate lead to immediate and intense seizure activity. There are continuous clonic seizures, followed by death. Seizure-resistant animals, injected in the same manner, start seizuring within 5 min, and death usually ensues within 3 days [Huxtable et al, Lippincott, 1979; Huxtable, 1980b]. This compound is worthy of further study by those capable of measuring physiological and behavioral dysfunctions.

Studies on the Mechanism of the Anticonvulsant Action of Taurine

The anticonvulsant action of taurine has been more thoroughly studied than its other CNS effects (for review see [Huxtable and Barbeau, 1976; Barbeau and Huxtable, 1978]). Table IX summarizes the various types of experimental seizures, on which taurine has been tested, and Table X summarizes clinical studies on the antiepileptic action of taurine. We have carried out a number of investigations of taurine and epilepsy since 1974, and I will discuss this aspect of the central nervous system pharmacology of taurine in greater detail to illustrate the difficulties of deducing physiological actions from pharmacological properties. Are there biochemical defects in epilepsy that involve taurine, and is the effectiveness of taurine due to some type of replacement therapy? Is the anticonvulsant action of taurine an exacerbation of a physiological effect, or is it unrelated to any of the normal roles of taurine in the brain?

Alterations in concentrations of amino acids in epileptic foci have been found in a number of experimental models (see, for example, review by [Emson, 1978]). GABA concentrations are lowered at the focus in cobalt-treated mice and cats, and in penicillin foci in cats. Freeze foci show no change. It has been

TABLE IX. Studies of the Antiepileptic Actions of Taurine

Type of epilepsy	Species	Route of taurine administration	Reference
Genetic-spontaneous	Cat	Subcutaneous/ oral	Van Gelder [1978b]
	Cat	Subcutaneous/ oral	Van Gelder et al [1977]
Genetic-photic	Baboon	Intraperitoneal	Wada et al [1975]
	Baboon	Intraperitoneal	Derouaux et al [1973]
Genetic-audiogenic	Rat	Intraperitoneal	Laird and Huxtable [1976][a]
Genetic-audiogenic	Rat	ICV	Laird and Huxtable [1976] Laird and Huxtable [1978] Huxtable and Laird [1978]
Genetic-intracerebral electroshock	Rat	Intra-inferior collicular	Laird and Huxtable [1978] Huxtable and Laird [1978]
Genetic-tossing	Mouse	Intraperitoneal	Iwata et al [1979]
Kindled	Rat	Intraperitoneal	Gaito [1976][a,d]
Kindled	Rat	Intraperitoneal	Burham et al [1975][a]
Kindled	Rat	ICV	Burham et al [1978][a]
Kindled-amygdaloid	Rat	Intraperitoneal	Wada et al [1975][a]
	Baboon	Intraperitoneal	Wada et al [1975][a]
	Cat	Intraperitoneal	Wada et al [1975][a]
Cobalt focus	Rat	Cortical superfusion	Mutani et al [1978]
		Cortical superfusion	Durelli et al [1976]
		Oral	Joseph and Emson [1976][a]
		Intraperitoneal	Joseph and Emson [1976][a]
		ICV	Joseph and Emson [1976][a]
	Cat	Subcutaneous	Van Gelder [1972]
		Subcutaneous	Van Gelder and Courtois, [1972]
		Subcutaneous	Van Gelder [1976]
		Superfusion	Durelli et al [1977]
		Intraarterial	Adembri et al [1974]
Cobalt focus	Mouse	Intraperitoneal	Van Gelder [1972] Carruthers-Jones and Van Gelder [1978]
	Rabbit		Savoldi et al [1976]

(continued next page)

TABLE IX (continued).

Type of epilepsy	Species	Route of taurine administration	Reference
Penicillin focus	Rat	Cortical superfusion	Mutani et al [1978]
Penicillin focus	Cat	Intravenous	Frigyesi and Lombardini [1979][c]
Penicillin focus	Cat	Cortical superfusion	Durelli et al [1976]
			Fariello et al [1975]
			Mutani et al [1977]
		Intravenous	Mutani et al [1974]
Aluminum focus			Mutani et al [1974]
Electroshock	Mouse	Intraperitoneal	Boggan et al [1978][a]
	Mouse		Thursby and Nevis [1974]
	Rat		Thursby and Nevis [1974]
Low calcium	Cat	Superfusion	Kaczmarek and Adey [1974]
Hyperbaric	Rat	Intraperitoneal	Adembri et al [1974][a,b]
Hypoxia	Rat	Intraperitoneal	Sandburg and Willow [1980]
Physostigmine	Cat	Intraarterial	Adembri et al [1974][a]
Monosodium glutamate	Rat	Intraperitoneal	Bradford and Dodd [1977]
Strychnine	Cat	Intraarterial	Adembri et al [1974][a]
		Intravenous	Mutani et al [1974]
	Rabbit	Intravenous	Roches et al [1979]
		Intravenous	Roches et al [1978]
Kynurenine	Mouse	Intraperitoneal	Lapin [1980]
Pentylene tetrazole	Mouse	ICV	Izumi et al [1975]
	Rat	Intraperitoneal	Bradford and Dodd [1977][a]
Ouabain (ICV)	Rat	ICV	Izumi et al [1973]
			Donaldson et al [1972]
			Tsukada et al [1974]
Estrogen	Cat	Intravenous	Mutani et al [1974]

[a]No anticonvulsant effect observed.
[b]Homotaurine protected against seizures.
[c]Augmentation of epilepsy by taurine.
[d]Onset of seizures delayed by taurine.

TABLE X. Clinical Trials of Taurine as an Antiepileptic Agent

Barbeau and Donaldson [1973]
Bergamini and Mutani [1973]
Barbeau and Donaldson [1974]
Bergamini et al [1974]
Sbarbaro [1974]
Striano and Buscaino [1974]
Bergamini and Mutani [1975][c]
Borromei et al [1975][b]
Marchesi et al [1975]
Mutani et al [1975]
Van Gelder et al [1975][a]
Barbeau et al [1976]
Mutani et al [1978]
Takahashi and Nakane [1978]
Mantovani and DeVivo [1979][d]

[a]No studies on clinical effects were performed.
[b]Quoted in Barbeau [1975].
[c]Quoted in Mutani et al [1978].
[d]No anticonvulsant effect observed.

fairly consistently observed that glutamate levels are lowered in all types of foci. Perry and co-workers [1975], however, found increases in human epileptic foci. Aspartate levels are decreased in cobalt-treated cats and rats, but not in mice or in freeze and penicillin foci. Glycine levels are elevated in cobalt-treated cortexes, but there are no changes in freeze or penicillin foci. The data for taurine are more confusing. Decreased levels have been found in cobalt-treated mice and rats, and in penicillin foci in cats, but in cobalt-treated cats levels are either unchanged or elevated. Others have reported no correlation between taurine levels and induction of penicillin foci in cats [Frigyesi and Lombardini, 1978]. No change is seen in freeze foci.

There has been a sharp difference of opinion concerning levels of amino acids in epileptic foci in humans. Perry and Hansen [1976] analyzed surgically removed focal tissue and compared the levels found with those in supposedly normal tissue removed of necessity during surgery. They found significant increases in taurine, glutamate, and GABA in epileptic foci compared to control cortical levels. At about the same time on the opposite side of the continent, Van Gelder [1976] had obtained antipodal results. In his case, he compared levels found in epileptic foci with levels in cortical brain tissue removed from patients undergoing surgery for reasons other than epilepsy. In the epileptic tissue, taurine and glutamate levels were significantly decreased and glycine levels significantly elevated. Thus, whereas Perry found a 42% increase in taurine, and 30% increase in glycine and GABA, and a 19% increase in glutamate, Van Gelder found a 15% decrease in taurine, an 11% decrease in glutamate, no change in GABA, and a 98% increase in glycine. Both sets of data are presumably accurate, but which set is the more meaningful has yet to be decided, although the work discussed in the rest of this section may give some indications.

The observation that in certain experimental models of epilepsy taurine and glutamate concentrations are both decreased does not necessarily imply that the loss of the one is causing the loss of the other [Van Gelder and Drujan, 1980]. Neurons maintain high concentrations of these amino acids relative to extracellular space, but appear to rely on outside sources for both of them—glutamate from the glial depots of glutamine, and taurine by direct transport. An increase in permeability of neurons would result in passive efflux of both taurine and glutamate. Under normal conditions, these substances would be captured by glia, which have higher transport capacities than do neurons for both amino acids. High rates of neuronal release, however, could in principle saturate glia transport systems leading to a net loss of the substances from the tissue.

In cobalt-treated cortical seizures in cats and mice a number of reports have indicated that oral taurine administration normalizes taurine, glutamate, and glycine levels in the epileptic tissue. One report failed to find any alteration in glutamate levels after taurine [Joseph and Emson, 1976]. Clearly, brain studies are difficult to do in humans. However, a number of trials have claimed that oral taurine administration either raises or normalizes serum glutamate levels in humans [Bergamini et al, 1974; Van Gelder et al, 1975; Takahashi and Nakane, 1978; Mutani et al, 1978]. On the basis of these and other findings, a unified hypothesis has been proposed to explain the biochemical bases of a number of types of epilepsy and to explain the anticonvulsant action of taurine. The theory is complex, and certain crucial steps in it have never been clearly explained, so I will give only a simplified description of it. The normal neuron has a high concentration of glutamate, whereas the normal glia cell has a high concentration of glutamine. The glutamate concentration is in a state of dynamic equilibrium, glutamate being interconverted with aspartate, formed from glutamine in the neuron, and converted irreversibly to GABA. The function of the glial pool of glutamine can be thought of as that of a reservoir to buffer glutamate concentration in the neuron. Excess glutamate is transported from the neuron to the glia and converted to glutamine, whereas any tendency to depress glutamate levels is countered by increased movement of glutamine from the glia to the neurons, and conversion to glutamate. Glutamate levels are normally much higher inside the neuron than in intercellular space. Anything that impairs the permeability barrier to glutamate, and allows neurons to lose this substance and extracellular levels to rise will predispose to seizures, due to the excitatory nature of glutamate. Van Gelder hypothesizes that the anticonvulsant action of taurine is due to its effect on increasing the movement of glutamine from glia to neuron. This pharmacological effect will be seen whether or not indigenous taurine levels are normal or depressed. Thus, the anticonvulsant action of taurine is secondary to a normalization or a raising of neuronal glutamate levels [Van Gelder, 1976, 1978a,b, 1980]. A considerable body of work on animal models of epilepsy appears to support this theory. A mechanism whereby taurine affects glutamate levels has not been suggested, except insofar as "A hyperexcitable condition is typified by a reduced ability to retain taurine and glutamic acid (cause and consequence?) . . ." [Van Gelder, 1978].

Most of the findings on which this hypothesis is based were obtained from cobalt-induced epilepsy. The vast majority of human epilepsies are genetic, and thus not closely related to the type of seizure that is produced when cobalt is placed on the cortex of an animal. Unlike a genetic epilepsy, in which the biochemical defect will occur wherever the gene is expressed, focal seizures, such as are produced in the cobalt model, involve disturbances in biochemistry at a discrete location, and the spread of a wave of excitation from there through normal brain tissue. In addition, anatomical changes are produced at the focus, such as areas of necrosis or gliosis. Such changes per se could be expected to lead to modifications of free amino acid concentrations. When modified levels are found, therefore, it is difficult to decide which changes are unconnected with the epileptogenicity, and which changes are involved in the epileptogenic process. There are also practical difficulties with such models. Seizures are spontaneous and beyond the experimenter's control, and the frequency of seizures varies with time. Thus, with cobalt, seizures start at about 2 days, and continue with decreasing frequency over a period of weeks. The dangers of deducing general conclusions on biochemical changes in epilepsy from studies on focal seizures have been stressed before [Emson, 1978]. In our own work, therefore, we chose to investigate the antiepileptic action of taurine on a genetic epilepsy. Genetic epilepsies occur in a large number of animals, including the mouse, rat, beagle, baboon, hamster, gerbil, chicken, and cat. We chose to use the seizure-susceptible rat (the so called "audiogenic" rat). Our earlier work with this model has been reviewed and summarized [Laird and Huxtable, 1978]. The term seizure-susceptible rat is preferred for these animals because they are, in fact, susceptible to a range of stimuli as well as loud noises. They require a lesser voltage to produce whole brain electroshock convulsions than do control animals, they require more anticonvulsants to protect against seizure activity, and chemoshock seizures are produced with lower doses of pentylene tetrazole (Table XI). Furthermore, intracerebral electroshock thresholds are lower than in control rats, both for nuclei along the auditory tract and outside of it (Table XII).

Among the experimental advantages of this epileptic model is that animals respond in a consistent manner to the same stimulus. In a given rat, sound will always induce a seizure of the same intensity, and intracerebral electroshock thresholds remain invariant over periods of months. Measurement of these thresholds permits precise quantification of seizure susceptibility. Furthermore, the seizures, not being spontaneous, are under the experimenter's control.

TABLE XI. Comparison of Seizure-susceptible and Seizure-resistant Rats

	Resistant	Susceptible
Electroshock thresholds (mA)		
Minimal Seizure[a]	25.9	16.2
Maximal Seizure[b]	66.0	23.5
Median anticonvulsant dose (mg/kg)[b]		
Tetrahydrocannabinol	16.0	35.0
Phenytoin	3.0	13.2
Carbamazepine	1.4	3.7
Phenobarbital	0.8	4.4
Median convulsant dose (mg/kg)[c]		
Pentylene tetrazole	19.5	11.2
Strychnine	0.42	0.39
Bicuculline	0.17	0.17

[a]Duplisse et al [1973].
[b]Laird et al [1975].
[c]Laird and Huxtable [1978].

TABLE XII. Intracerebral Electroshock Thresholds of Seizure-susceptible and Seizure-resistant Rats*

Brain	Resistant (μA)	Susceptible (μA)	Change (%)
Inferior colliculus	188.6	45.9	24
Medial geniculate body	463.0	92.5	20
Ventral Cochlear Nucleus	>1000	45.0	<4
Reticular formation	188.8	95.2	50

*Data from Huxtable and Laird [1978]. Animals were implanted bilaterally with bipolar electrodes, and thresholds represent the minimum current required to cause convulsions.

In order to study the antiepileptic action of taurine in this model, we must first ensure that taurine has the protective effect on genetic seizures that it has on pharmacological seizures, such as ouabain, cobalt, or freeze induced. When this has been established, if a biochemical change associated with taurine is seen with focal but not genetic seizures, we may assume that the change is not connected with the antiepileptic action of taurine.

Does taurine attenuate seizure-susceptibility in genetically seizure-susceptible rats? Two parameters are used to quantitate an anticonvulsant action: change in threshold, which is an indication of how resistant the brain is to seizure initiation; and change in seizure severity, which is an indication of how readily electrical discharge will spread through the brain after a focus is initiated. Seizure severity can be measured without regard to the experimental stimulus used to cause a seizure. Electrical thresholds, of course, may only be measured for seizures induced by intracerebral electrical stimulation.

Taurine, given ICV, decreased the severity of sound-induced seizures in these animals in a dose-related manner (Fig. 7). In this regard, it is considerably more potent than GABA, glycine, or aminoisobutyric acid (a transportable but non-metabolizable amino acid). Given by this route, taurine also markedly attenuated the severity of seizures elicited by electrical stimulation of the inferior colliculus. This attenuation was observed within 3 min of injection and persisted for 24 hr. The electrical threshold for seizure showed a biphasic response. A significant elevation of threshold was seen within 3 min, but by 24 hr this returned to control level. At 48 hr, the threshold was again significantly elevated, and remained elevated through the sixth day. When the same doses were given of radioactive taurine, it could be calculated from the radioactivity reaching the inferior colliculus within 3 min that 200 nmole of taurine had reached this structure. If this quantity of taurine was injected directly into the inferior colliculus, no effect on seizure intensity was observed. However, within 12 min the threshold for seizure was significantly elevated, from 50 to 80 μA. By 24 hr, the threshold had reached 95 μA, and remained significantly elevated above control through the 18th day

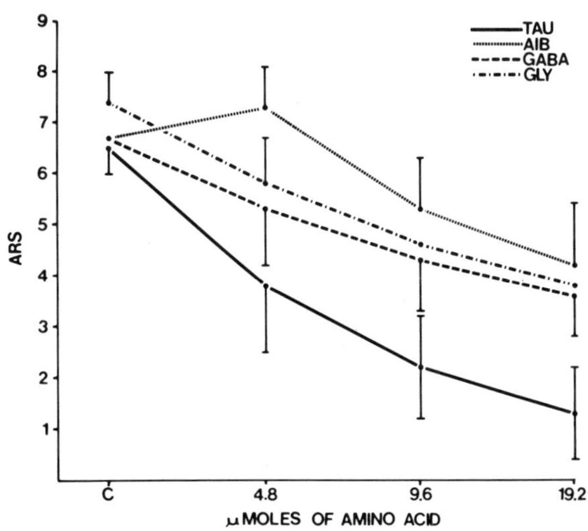

Fig. 7. Effect of amino acids given ICV on sound-induced seizure intensity. Each point is mean ± SE of six animals. All doses of taurine caused significant reductions (P<0.01). The highest doses of the other amino acids and the next highest dose of GABA and glycine were significant at the P<0.05 level. Data from Huxtable and Laird [1978].

(Fig. 8). GABA showed a similar phenomenon, the maximum elevation in threshold being less, and the duration of action shorter. These long-lasting anticonvulsant actions following a single injection are unprecedented for taurine.

This model of epilepsy, therefore, reproduces the pharmacological effects of taurine that have been observed on the cobalt model and other models. Having established this, our attention was directed toward finding what indigenous differences in biochemistry these animals had, and what biochemical effects taurine was having that could account for the antiepileptic action. This has resulted in a long series of negative findings. Areas showing decreased thresholds for seizures, such as the inferior colliculus, had no alteration in free amino acid levels when compared to seizure-resistant animals [Huxtable and Laird, 1978b]. Furthermore, none of the amino acid ratios reported to be disturbed in cobalt-induced epilepsy were altered in seizure-susceptible animals. These findings point to a gross difference in biochemistry between this model and other pharmacological models. Although taurine is anticonvulsant in both types of epilepsy, it can not be acting in the seizure-susceptible rat by rectification of abnormal glutamate levels as has been proposed for cobalt epilepsy, because glutamate levels are not abnormal. Furthermore, when amino acid levels are determined in the inferior colliculus 24 hr after an injection of taurine, ie, at the time of maximum raising of threshold, no significant alteration in amino acid levels can be found (Table XIII).

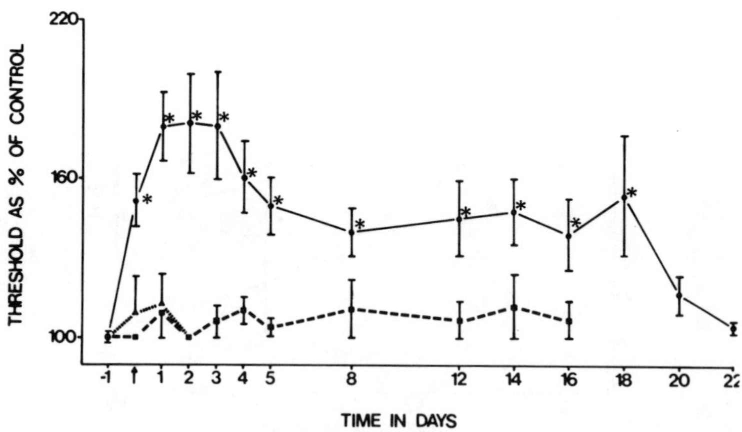

Fig. 8. Time course of anticonvulsant action of taurine. Taurine (200 nmol) was injected bilaterally into the inferior colliculi. Control thresholds were determined 24 hr prior to injection. Arrow is 12 min following injection. Data are means ± SE of 10 animals. * P<.05 ····· Saline-injected controls.---Taurine-injected seizure-resistant animals. Data from Laird and Huxtable [1978].

TABLE XIII. Amino Acid Concentrations in Inferior Colliculus of Seizure-susceptible Rats After Unilateral Injection of Taurine*

Amino acid	Noninjected side µmol/g wet weight	Injected side µmol/g wet weight
Phosphoethanolamine	0.66 ± 0.32	0.75 ± 0.49
Taurine	1.92 ± 0.52	2.15 ± 0.60
Threonine	0.45 ± 0.16	0.47 ± 0.17
Serine	0.47 ± 0.13	0.46 ± 0.19
Aspartate	2.98 ± 1.34	3.44 ± 1.10
Glutamine	2.05 ± 0.81	2.07 ± 0.57
Glutamate	5.22 ± 2.58	5.70 ± 2.06
Glycine	1.45 ± 0.44	1.52 ± 0.36
Alanine	0.78 ± 0.44	0.83 ± 0.78
GABA	3.65 ± 1.70	3.70 ± 1.65

*Taurine (200 nmol) was injected unilaterally, and the animals were sacrificed 24 hr later, at a time when seizure thresholds were still raised as a result of the injection. Data are means ± SD of nine animals. [Laird and Huxtable, unpublished].

Brain amino acid concentrations have also been found to be normal in the genetically photoepileptic baboon [Hansen et al, 1973].

Glutamic acid decarboxylase activities were no different from control when measured both in the presence and absence of pyridoxal 5′-phosphate in cortex, corpus striatum, and inferior colliculus. No difference could be observed in the rate of loss of [^{3}H]taurine from the inferior colliculus of seizure-susceptible and seizure-resistant animals. However, after peripheral injection of [^{3}H]taurine, seizure-susceptible rats had a lower uptake into the inferior colliculus, reticular formation, frontal cortex, and cerebellum. The inferior colliculus had the lowest uptake relative to the control (61% lower). Defects in transport of taurine have also been observed in a number of other situations. Taurine uptake by platelets is significantly decreased in myoclonus epilepsy [Airaksinen, 1979], and kidney resorption is impaired in Freidreich's ataxia [Filla et al, 1979]. On the other hand, taurine transport is elevated in genetic cardiomyopathy in hamsters [Azari and Huxtable, 1980b].

Catecholamine balance is markedly different in seizure-susceptible animals [Laird et al, 1980]. Norepinephrine content is significantly lowered in inferior colliculus, cerebral hemisphere, midbrain, hypothalamus, and pons medulla of seizure-susceptible rats, but levels in the cerebellum are normal. Serotonin content is significantly decreased in cerebral hemispheres and midbrain.

In the inferior colliculus, norepinephrine levels are 27% lower in seizure-susceptible animals relative to resistant animals. Twelve minutes after injection into the inferior colliculus of 200 nmol of taurine (ie, when the anticonvulsant action is first manifested), norepinephrine levels have increased 17% [Laird, unpublished data]. This is the first finding of a biochemical effect of taurine. These observations give us a basis for further investigations of the biochemical basis of this genetic epilepsy and the mechanism of the anticonvulsant action of taurine.

Van Gelder's theory concerning the anticonvulsant effect of taurine is clearly not valid for the seizure-susceptible rat. This may mean that the mechanism of action of taurine is different in this model of epilepsy than in the ones in which amino acid disturbances are found. Or, it may mean that changes in amino acid levels are parallel phenomenon not related to the anticonvulsant effect. The similarity in pharmacological action of taurine or the different types of epilepsy strongly suggests a similar mechanism, although the data at this time do not prove or disapprove this. The seizure-susceptible rat is probably a better model for clinical epilepsies than are nongenetic laboratory models. Van Gelder's unifying theory has been highly fruitful in stimulating work in this area over a number of years. The theory may have validity for some epilepsies, but it certainly can not be held as a universal theory of epilepsy.

ACKNOWLEDGMENTS

Work quoted herein from my laboratory was supported by USPHS Grants HL 19394 and NS 14405. All the studies on epilepsy were performed in close collaboration with Dr. Hugh E. Laird, University of Arizona.

REFERENCES

Major reviews on taurine may be found in: Jacobsen and Smith [1968]; Huxtable and Barbeau [1976]; Oja, Kontro, and Lähdesmäki [1977]; Barbeau and Huxtable [1978]; Cavallini, Gaull, and Zappia [1980]; and Huxtable and Kuriyama, [1980].

1. Adembri G, Bartolini A, Bartolini R, Giotti A, Zilletti L: Anticonvulsive action of homotaurine and taurine. Br J Pharmacol 52:439–440, 1974.
2. Agrawal HC, Davison AN, Kaczmarek LK: Subcellular distribution of taurine and cysteine-sulphinate decarboxylase in developing rat brain. Biochem J 122:759–763, 1971.
3. Agrawal HC, Himwich WA: Amino acids, proteins and monoamines of developing brain. In: Himwich WA (ed): "Developmental Neurobiology." Springfield, Ill: C.C. Thomas, 1970, pp 287–310.
4. Airaksinen EM: Uptake of taurine, GABA, 5-HT and dopamine by blood platelets in progressive myoclonus epilepsy. Epilepsia 20:503–510, 1979.
5. Anderson EG: Are there too many neurotransmitters? Trends in Pharmacology 1:246–247, 1980.
6. Anderson G, Ashley VM, Jones D: Utilization of L-methionine sulfoxide, L-methionine sulfone and cysteic acid by the weanling rat. J Nutr 106:1108–1114, 1976.
7. Austin L, Recasens M, Mathur RL, Mandel P: The distribution of taurine, cysteinyl sulphinilic acid decarboxylase and crysteinyl (sic) sulphinilic acid α-ketoglutarate transaminase in the rat olfactory bulb and ofactory nucleus. Neuroscience Lett 9:59–63, 1978.
8. Awapara J, Manz N: Absorption of injected taurine ^{35}S by rat organs. J Biol Chem 225:877–882, 1957.

9. Azari J, Bahl J, Huxtable R: Guanidinoethyl sulfonate and other inhibitors of the taurine transport system in the heart. Proc W Pharmocol Soc 22:289–393, 1979.
10. Azari J, Brumbaugh P, Huxtable R: Prophylaxis by taurine in the hearts of cardiomyopathic hamsters. J Mol Cell Cardiol 122:1353–1366, 1980.
11. Azari J, Huxtable RJ: The mechanism of the adrenergic stimlation of taurine influx in the heart. Eur J Pharmacol 61:217–223, 1980.
12. Balcar V, Borg J, Robert J, Mandel P: Uptake of L-glutamate and taurine in neuroblastoma cells with altered fatty acid composition of membrane phospholipids. J Neurochem 34:1678, 1980.
13. Barbeau A: The brain, the heart and taurine. Canad J Neurol Sci 2:343–346, 1975.
14. Barbeau A, Donaldson J: Taurine in epilepsy. The Lancet II:387, 1973.
15. Barbeau A, Donaldson J: Zinc taurine and epilepsy. Arch Neurol 30:52–58, 1974.
16. Barbeau A, Huxtable R: "Taurine and Neurological Disorders." New York: Raven Press, 1978.
17. Barbeau A, Tsukada Y, Inoue I: Neuropharmacology of taurine. In Huxtable R, Barbeau A (eds): "Taurine." New York: Raven Press, 1976, pp 253–263.
18. Bender DA: "Amino Acid Metabolism." New York: Wiley, 1975.
19. Benjamin AM, Quastel JH: Effects of acetylcholine on potassium-induced changes of GABA and taurine uptakes and release in cerebral cortex slices from the rat. Canad J Physiol Pharmacol 55:356–362, 1977.
20. Bergamini L, Mutani R: L'action anti-epileptique de la taurine chez l'homme. Le Concours Medical (Paris) 35:25–28, 1975.
21. Bergamini L, Mutani R: Taurine: A new approach to treatment of epilepsy. Rev N Sper Freniat 97:783–787, 1973.
22. Bergamini L, Mutani R, Delsedime M, Durelli L: First clinical experience on the antiepileptic action of taurine. Eur Neurol 11:261, 1974.
23. Bergeret B, Chatagner F: Sur la prescence d'acide cysteine sulfinique dans le cerveau du rat normal. Biochim Biophys Acta 14:297, 1954.
24. Blake RL, Grillo RV, Russell, ES: Increased taurine excretion in hereditary hyperprolinemia of the mouse. Life Sci 14:1285–1290, 1974.
25. Boggan WO, Medberry D, Hopkins DH: Effect of taurine on some pharmacological properties of ethanol. Pharmacol Biochem Behav 9:469–472, 1978.
26. Borg J, Balcar VJ, Mandel P: High affinity uptake of taurine in neuronal and glial cells. Brain Res 118:514–516, 1976.
27. Borg J, Ramaharobandro N, Mark J, Mandel P: Changes in the uptake of GABA and taurine during neuronal and glial maturation. J Neurochem 34:1113–1122, 1980.
28. Borromei A, Sacquegna T, Coccagna G: Use of taurine in continuous partial epilepsy. Observations of a case. Phronesis 17:401–415, 1975.
29. Bradford HF, Dodd PR: Convulsions and activation of epileptic foci induced by monosodium glutamate and related compounds. Biochem Pharmacol 26:253–254, 1977.
30. Burnham WM, Albright, PA, Racine RJ: The effect of taurine on kindled seizures in the rat. Canad J Physiol Pharmacol 56:497–500, 1978.
31. Carruthers-Jones DI, Van Gelder NM: Influence of taurine dosage on cobalt epilepsy in mice. Neurochem Res 3:115–123, 1978.
32. Cavallini D, Gaull GE, Zappia V: "Natural Sulfur Compounds: Novel Biochemical and Structural Aspects." New York: Plenum, 1980.
33. Cavallini D, Scandurra R, Dupre S: Cysteamine oxygenase (horse kidney). Method Enzymol XVIIB:479–483, 1971.
34. Chanda R, Himwich WA: Taurine levels in developing rabbit brain and other organs. Develop Psychobiol 3:191–196, 1970.
35. Chasov R, Mal'chikova L, Lipina N, Reshetnaya N, Asafov G, Smirnov V: Taurine and the electrical activity of the heart. Circ Res 35:III-II-III-21, 1974.

36. Chesney RW, Jax DK, Scriver CR, Mohyuddin F: Taurine transport in mammalian kidney. In Barbeau A, Huxtable R (eds): "Taurine and Neurological Disorders." New York: Raven Press, 1978, pp 73–93.
37. Chesney RW, Scriver CR, Mohyuddin F: Localization of the membrane defect in transepithelial transport of taurine by parallel studies in vivo and in vitro in hypertaurinuric mice. J Clin Invest 57:183–193, 1976.
38. Chubb J, Huxtable R: Isoproterenol-stimulated taurine influx in the perfused rat heart. Eur J Pharmacol 48:369–376, 1978.
39. Close R: Free amino acids of some fungi. Nature (London) 185:609, 1960.
40. Collins GGS: The rates of synthesis, uptake, and disappearance of (^{14}C) taurine in eight areas of the rat central nervous system. Brain Res 76:447–459, 1974.
41. Cook DA, Hagerman LM, Schneider DL, Sarett HP: Effect of dietary taurine on fecal bile salt excretion in rats and hamsters fed cholestryramine. Proc Soc Exp Biol Med 138:830–834, 1971.
42. Davis JM, Himwich WA: Amino acids and proteins of developing mammalian brain. In Himwich W (ed): "Biochemistry of the Developing Brain." New York: Dekker, 1973, pp 55–110.
43. Derouaux M, Puil E, Naquet R: Antiepileptic effect of taurine in photosensitive epilepsy. Electroencephologaph Clin Neurophysiol 35:770, 1973.
44. DiGiorgi RM, Tucci G, Macaione S: Cysteine oxidase activity in rat retina during development. Life Sci 16:429–435, 1975.
45. Donaldson J, Minnich JL, Barbeau A: Ouabain-induced seizures in rats: Regional and subcellular localization of ^{3}H-ouabain associated with NA+ K+-ATPase in brain. Canad J Biolchem 50:888–896, 1972.
46. Duplisse BR, Picchioni AL, Chin L, Consroe PF: Electroshock seizure thresholds: Differences between audiogenic and non-audiogenic rats. J Int Res Commun 1:9, 1973.
47. Durelli L, Mutani R, Delsedime M, Quattrocolo G, Buffa C, Mazzarino M, Fumero S: Electroencephalographic and biochemical study of the antiepileptic action of taurine administered by cortical superfusion. Exp Neurol 52:30–39, 1976.
48. Durelli L, Mutani R, Quattrocolo G, Delsedime M, Buffa C, Fassio F, Valentini C, Fumero S: Relationships between electroencephalographic pattern and biochemical picture of the cobalt epileptogenic lesion after cortical superfusion with taurine. Exp Neurol 54:489–503, 1977.
49. Dziewiatkowski DD: Utilization of sulfate sulphur in the rat for the synthesis of cysteine. J Biol Chem 207:181–186, 1954.
50. Emson PC: Biochemical and metabolic changes in epilepsy. In Barbeau A, Huxtable R (eds): "Taurine and Neurological Disorders." New York: Raven Press, 1978, pp 319–338.
51. Ennor AH, Morrison JF: Biochemistry of the phosphagens and related guanidines. Physiol Rev 38:631–674, 1958.
52. Ericson LE, Carlson B: Studies on the occurrence of amino acids, niacin and pantothenic acid in marine algae. Arkiv Kemi 6:511–522, 1954.
53. Ewetz L, Sorbo B: Characteristics of the cysteine sulfinate-forming enzyme system in rat liver. Biochim Biophys Acta 128:296–305, 1966.
54. Fariello RG, Lloyd KG, Hornykievicz O: Cortical and sub-cortical projected foci in cats: Inhibitory action of taurine. Neurology 25:1077–1083, 1975.
55. Federici G, Dupre S, Rosei MA, Granta F, Orlando M: The metabolism of taurine in the living rat. Physiol Chem Phys 6:411–416, 1974.
56. Fellman JH, Roth ES, Fujita TS: Taurine is not metabolized to isethionate in mammalian tissue. In Barbeau A, Huxtable R (eds): "Taurine and Neurological Disorders." New York: Raven Press, 1978, pp 19–24.
57. Filla A, Butterworth RF, Barbeau A: Pilot studies on membranes and some transport mech-

anisms in Friedreich's ataxia. Canad J Neurol Sci 6:285–289, 1979.
58. Fiori A, Costa M: Oxidation of hypotaurine by H_2O_2. Acta Vitaminol Enzymol 23:204–207, 1969.
59. Frigyesi TL, Lombardini JB: Lack of correlation between taurine levels in 16 brain regions and paroxysmal discharges in the thalamocortical circuit. Neurosci Lett 7:213–217, 1978.
60. Frigyesi TL, Lombardini JB: Augmentation of thalamomotor cortico cerebellar epileptogenesis by taurine and its antagonism by diphenyl hydantoin. Life Sci 24:1251–1260, 1979.
61. Fromageot PF, Chapeville F: Incorporation de sulfite dans l'animal superieur. In "Colloque sur la Biochimie du Soufre, Roscoff." Paris: Centre National de la Recherche Scientifique, 1956, pp 171–180.
62. Fugelli K, Zachariassen KE: The distribution of taurine, GABA, and inorganic ions between plasma and erythrocytes in flounder (Platichthys flesus) at different plasma osmolalities. Comp Biochem Physiol 55A:173–177, 1976.
63. Gabellec MM, Recasens M, Mandel P: Developmental pattern of cysteine sulphinic acid transaminase activity in some areas of mice nervous system. Life Sci 23:1263–1269, 1978.
64. Gaito J: The effects of taurine on various stages of the kindling process. A summary of results. Bull Psychon Soc 7:397–400, 1976.
65. Gaull GE, Rassin DK: Taurine and brain development: Human and animal correlates. In Meisami E, Brazier MAB (eds): "Neural Growth and Differentiation." New York: Raven Press, 1979, pp 461–477.
66. Gaull GE, Rassin DK: Taurine in development and nutrition. In "Sulphur in Biology," Amsterdam: Excerpta Medica, 1980, pp 271–288.
67. Gaull GE, Rassin DK, Raiha NCR, Heinonen K: Milk protein quantity and quality in low birth weight infants. III. Effects on sulfur-containing amino acids in plasma and urine. J Pediatr 90:348–355, 1977.
68. Green JP, Robinson JD: Cerebroside sulfate (Sulfatide A) in some organs of the rat and in most mouse cells. J Biol Bhem 235:1621–1624, 1960.
69. Guidotti A, Badianti G, Pepeu G: Taurine distribution in cat brain. J Neurochem 19:431–435, 1972.
70. Hansen S, Perry TL, Wada JA, Sokol M: Brain amino acids in baboons with light-induced epilepsy. Brain Res 50:480–483, 1973.
71. Haslewood GAD: Bile salts: Structure, distribution, and possible biological significance as a species character. In Florkin M, Mason HS (eds): "Comparative Biochemistry." Vol. 3, New York: Academic Press, 1962, pp 205–229.
72. Hill LJ, Martin WG: The synthesis of taurine from sulfate V. Regulatory modifiers of the chick liver enzyme system. Proc Soc Exp Biol Med 144:530–533, 1973.
73. Himwich WA, Agrawal HC: Amino acids. Handbook of Neurochemistry 1:33–52, 1969.
74. Honegger CG, Krepelka LM, Steiner M, Hahn HP von: Kinetics and subcellular distribution of S^{35}-taurine uptake in rat cerebral cortex slices. Experientia 29:1235–1237, 1973.
75. Hoskin FCG, Brande M: An improved sulphur assay applied to a problem of isethionate metabolism in squid axon and other nerves. J Neurochem 20:1317–1327, 1973.
76. Howard AM, Courtenay Evans RJ: Secholex, clofibrate and taurine in hyperlipidemia. Atherosclerosis 20:105–116, 1974.
77. Hoyaux J, Gilles R, Jeuniaux CH: Osmoregulation in molluscs of the intertidal zone. Comp Biochem Physiol 53A:361–365, 1976.
78. Hruska RE, Huxtable RJ, Yamamura HI: High-affinity, temperature-sensitive, and sodium-dependent transport of taurine in rat brain. In Barbeau A, Huxtable R (eds): "Taurine and Neurological Disorders." New York: Raven Press, 1978a, pp 109–117.
79. Hruska R, Padjen A, Bressler R, Yamamura HI: Taurine: Sodium-dependent, high-affinity transport into rat brain synaptosomes. Molec Pharmacol 14:77–85, 1978b.

80. Hruska RE, Thut PD, Huxtable RJ, Bressler R: Suppression of conditioned drinking by taurine and related compounds. Pharmacol Biochem Behav 3:593–599, 1975.
81. Huovinen JA, Gustafsson BE: Utilization of inorganic sulphate, sulphite and sulphide for sulphur amino acid biosynthesis in germfree and conventional rats. Acta Physiol Scand 68:Suppl 277:77, 1966.
82. Huxtable RJ: Metabolism and function of taurine in the heart. In Huxtable R, Barbeau A (eds): "Taurine." New York: Raven Press, 1976, pp 99–120.
83. Huxtable RJ: Regulation of taurine in the heart. In Barbeau A, Huxtable R (eds): "Taurine and Neurological Disorders." New York: Raven Press, 1978, pp 5–17.
84. Huxtable RJ: Does taurine have a function? Trends Pharmacol 1:6–7, 1979.
85. Huxtable RJ: Does taurine have a function? Introduction. Fed Proc 39:2678–2679, 1980a.
86. Huxtable RJ: The regulation and function of taurine in the heart and other organs. In Cavallini D, Gaull GE, Zappia V (eds): "Natural Sulfur Compounds: Novel Biochemical and Structural Aspects." New York: Plenum, 1980b, pp 277–293.
87. Huxtable RJ: Sources and turnover rates of taurine in nursing and weaned rat pups. J Nutrit (in press) 1981.
88. Huxtable R, Barbeau A: "Taurine." New York: Raven Press, 1976.
89. Huxtable R, Bressler R: Taurine and isethionic acid: Distribution and interconversion in the rat. J Nutrit 102:805–814, 1972.
90. Huxtable R, Bressler R: Taurine concentration in congestive heart failure. Science 184:1187–1188, 1974a.
91. Huxtable R, Bressler R: Elevation of taurine in human congestive heart failure. Life Sci 14:1353–1359, 1974b.
92. Huxtable R, Chubb J: Adrenergic stimulation of taurine transport by the heart. Science 198:409–411, 1977.
93. Huxtable RJ, Chubb J, Azari J: Physiological and experimental regulation of taurine content in the heart. Fed Proc 39:2685–2690, 1980.
94. Huxtable R, Kuriyama K: Does taurine have a function? Fed Proc 39:2678–2708, 1980.
95. Huxtable R, Laird H: The prolonged anticonvulsant action of taurine on genetically determined seizure susceptibility. Canad J Neurol Sci 5:215–221, 1978.
96. Huxtable RJ, Laird HE: Are amino acid levels necessarily abnormal in epileptic brains? Studies on the genetically seizure-susceptible rat. Neurosci Lett 10:341–345, 1978.
97. Huxtable RJ, Laird HE, Lippincott SE: The transport of taurine in the heart and the rapid depletion of tissue taurine content by guanidinoethyl sulfonate. J Pharmacol Exp Therap 211:465–471, 1979.
98. Huxtable R, Lippincott SE: Comparative metabolism and taurine depleting effects of guanidinoethyl sulfonate in cats, guinea pigs, and mice. Arch Biochem Biophys (in press), 1981.
99. Iwata H, Yamagami S, Lee E, Matsuda T, Baba A: Increase of brain taurine contents of E1 mice by physiological stimulation. Japan J Pharmacol 29:503–507, 1979.
100. Izumi K, Donaldson J, Munnich J, Barbeau A: Ouabain-induced seizures in rats: Suppressive effects of taurine and gamma-aminobutyric acid. Canad J Pharmacol 51:885–889, 1973.
101. Izumi K, Igishu H, Fukuda T: Suppression of seizures by taurine specific or non-specific? Brain Res 76:171–173, 1974.
102. Izumi K, Igisu H, Fukuda K: Effect of edetate on seizure suppressing actions of taurine and GABA. Brain Res 88:576–579, 1975.
103. Jacobsen JG, Collins LL, Smith LH: Urinary excretion of isethionic acid in man. Nature (London) 214:1247–1248, 1967.
104. Jacobsen JG, Smith LH: Comparison of decarboxylase of cysteine sulphinic acid-1-^{14}C and cysteic acid-1-^{14}C by human, dog, and rat liver and brain. Nature (London) 200:575–577, 1963.

105. Jacobsen JG, Smith LH: Biochemistry and physiology of taurine and taurine derivatives. Physiol Rev 48:424–511, 1968.
106. Jacobsen JG, Thomas LL, Smith LH: Properties and distribution of mammalian L-cysteine sulfinate carboxy-lyases. Biochim Biophys Acta 85:103–116, 1964.
107. Jakubke HD, Jeschkeit H: "Amino Acids, Peptides and Proteins." New York: Wiley, 1977.
108. Johnston GAR: Biochemistry of glycine, taurine, glutamate and asparate. In Iverson LL, Iverson SD, Synder SH (eds): "Handbook of Psychopharmacology Amino Acid Neurotransmitters." Vol 4, New York: Plenum, 1975, pp 59–82.
109. Joseph MH, Emson PC: Taurine and cobalt induced epilepsy in the rat: A biochemical and electrocorticographic study. J Neurochem 27:1495–1501, 1976.
110. Kaczmarek LK: A comparison of the evidence for taurine and GABA as neurotransmitters. In Huxtable R, Barbeau A (eds): "Taurine." New York: Raven Press, 1976, pp 283–292.
111. Kaczmarek LK, Adey LK, Adey WR: Factors affecting the release of (^{14}C) taurine from cat brain: The electrical effects of taurine on normal and seizure prone cortex. Brain Res 76:83–94, 1974.
112. Kaczmarek LK, Agrawal HC, Davison AN: The biochemistry of taurine in developing rat brain. In Carsen NAJ, Raine DN (eds): "Inherited Disorders of Sulphur Metabolism." London: Churchill, 1971, pp 63–68.
113. Kaczmarek LK, Davison AN: The physiological role of taurine. Biochem J 123:45P–46P, 1971.
114. Kacmarek LK, Davison AN: Uptake and release of taurine from rat brain slices. J Neurochem 19:2355–2362, 1972.
115. Knopf K, Sturman J, Armstrong M, Hayes KC: Taurine: An essential nutrient for the cat. J Nutrit 108:773–778, 1978.
116. Koechlin BA: On the chemical composition of the axoplasm of squid giant nerve fibers with particular reference to its ion pattern. J Biophys Biochem Cytot 1:511, 1955.
117. Kontro P: Transport and metabolism of taurine and hypotaurine in the brain. Thesis. University of Oulu, 1980.
118. Kontro P, Oja SS: Hypotaurine oxidation by mouse tissues. In Cavallini D, Gaull GE, Zappia V (eds): "Natural Sulfur Compounds." New York: Plenum Press, 1980, pp 201–212.
119. Lähdesmäki P, Oja SS: Effect of electrical stimulation on the influx and efflux of taurine in brain slices of newborn and adult rats. Exp Brain Res 15:430–438, 1972.
120. Laird HE, Chin L, Picchioni AL: Lower endogenous levels of norepinephrine and serotonin in the brains of rats from a genetically seizure-susceptible strain. Neurosci Abstr 6, 1980.
121. Laird H, Huxtable R: Effect of taurine on audiogenic seizure response in rats. In Huxtable R, Barbeau A (eds): "Taurine." New York: Raven, 1976, 267–274.
122. Laird H, Huxtable R: Taurine and audiogenic epilepsy. In Barbeau A, Huxtable R (eds): "Taurine and Neurological Disorder." New York: Raven, 1978, pp 339–357.
123. Laird HE, Huxtable R, Jones BC, Chin LC: Suppression of audiogenic seizure in rats by taurine, gamma-aminobutyric acid, glycine and aminoisobutyric acid. Neurosci Abstr 1:709, 1975.
124. Laird H, Lippincott S, Huxtable R: Contribution of dietary taurine to total body taurine in the rat. Fed Proc 39:1045, 1980.
125. Lange R: The osmotic function of amino acids and taurine in the mussel, *Mytitus edulis*. Comp Biochem Physiol 13:205–216, 1964.
126. Lange R, Fugelli K: The osmotic adjustment in the euryhaline teleosts, the flounder, *Pleuronectes flesus* L. and the three-spined stickleback, *Gasterosteus aculeatus* L. Comp Biochem Physiol 15:283–292, 1965.
127. Lapin IP: Taurine selectively antagonizes L-kynurenine-produced seizures in mice. J Neurol Transmission 48:311–316, 1980.

128. Lefauconnier J-M, Urban F, Mandel P: Taurine transport into the brain in the rat. Biochimie 60:381–387, 1978.
129. Lemieux B, Barbeau A, Beroniade V, Shapcott D, Breton G, Geoffroy G, Melancon S: Amino acid metabolism in Friedreich's ataxia. Canad J Neurol Sci 3:373–378, 1976.
130. Lin YC, Demeio RH, Metione RM: Purification and properties of rat liver cysteine sulfinate decarboxylase. Biochim Biophys Acta 250:558–567, 1971.
131. Lombardini JB: Regional and subcellular studies on taurine in the rat central nervous system. In Huxtable R, Barbeau A (eds): "Taurine." New York: Raven, 1976, pp 311–326.
132. Lombardini JB: High-affinity transport of taurine in the mammalian central nervous system. In Barbeau A, Huxtable R (eds): "Taurine and Neurological Disorders." New York: Raven, 1978, pp 119–136.
133. Lombardini JB, Homan JA: Rate of disappearance of isethionic acid in the rat central nervous system. Neurochem Res 4:449–455, 1979.
134. Mantovani J, DeVivo DC: Effects of taurine on seizures and growth hormone release in epileptic patients. Arch Neurol 36:672–674, 1979.
135. Marchesi G, Angeleri G, Quattrini A: Clinical and EEG study of the effects of high doses of taurine in epileptics: Preliminary investigation. Riv Neurol 45:123–127, 1975.
136. Marquardt P, Vogg G: Pharmakologische und chemische unterschungen uber wirkstoffe in bienenpollen. Arzneimittel Forsch 2:267–271, 1952.
137. Martin WG, Sass NL, Hill L, Tarka S, Truex R: The synthesis of taurine from sulfate IV. An alternative pathway for taurine synthesis by the rat. Proc Soc Exp Biol Med 141:632–633, 1972.
138. Martin WG, Truex CR, Tarka SM, Hill LJ, Gorby WG: The synthesis of taurine from sulfate VIII. A constitutive enzyme in mammals. Proc Soc Exp Biol Med 147:563–565, 1974.
139. Miraglia RJ, Martin WG, Spaeth DG, Patrick H: On the synthesis of taurine from sulfate by the chick. 1. Influential dietary factors. Proc Soc Exp Biol Med 123:725–730, 1966.
140. Misra CH, Olney JW: Cysteine oxidase in brain. Brain Res 97:117–126, 1975.
141. Mutani R, Bergamini L, Delsedime M, Durelli L: Effects of taurine in chronic experimental epilepsy. Brain Res 79:330–332, 1974.
142. Mutani R, Bergamini L, Durelli L: Taurine in experimental and human epilepsy. In Barbeau A, Huxtable R (eds): "Taurine and Neurological Disorders." New York: Raven Press, 1978, pp 359–374.
143. Mutani R, Bergamini L, Fariello R, Delsedime M: Effects of taurine on cortical acute epileptic foci. Brain Res 70:170–173, 1974.
144. Mutani R, Durelli L, Mazzarino M, Valentini C, Monaco F, Fumero S, Mondini A: Longitutidinal changes of brain amino acid content occurring before, during and after epileptic activity. Brain Res 122:513–521, 1977.
145. Mutani R, Monaco F, Durelli L, Delsedime M: Levels of free amino acids in serum and cerebrospinal fluid after administration of taurine to epileptic and normal subjects. Epilepsia 16:765–769, 1975.
146. Oja SS: Exchange of taurine in brain slices of adult and 7-day-old rats. J Neurochem 18:1847–1852, 1971.
147. Oja SS, Karvonen ML, Lähdesmäki P: Biosynthesis of taurine and enhancement of decarboxylation of cysteine sulphinate and gluatmate by the electrical stimulation of rat brain slices. Brain Res 55:173–178, 1973.
148. Oja SS, Kontro P, Lähdesmäki P: Amino acids as inhibitory neurotransmitters. Prog Pharamcol 1:1–119, 1977.
149. Pasantes-Morales H, Chatagner F, Mandel P: Synthesis of taurine in rat liver and brain in vivo. Neurochem Res 5:441–451, 1980.
150. Pasantes-Morales H, Loriette C, Chatagner F: Regional and subcellular distribution of taurine synthesizing enzymes in the rat central nervous tissue. Neurochem Res 2:671–680, 1977.

151. Pearson PB: Utilisation du soufre mineral par les animaux superieurs pour la sunthese de molecules organiques. In "Colloque sur la Biochimie du Soufre." Paris: CNRS, 1956, pp 153–170.
152. Peck Jr EJ, Awapara J: Formation of taurine and isethionic acid in rat brain. Biochim Biophys Acta 141:499–506, 1967.
153. Perry TL, Berry K, Hansen S, Diamond S, Mok C: Regional distribution of amino acids in human brain obtained at autopsy. J Neurochem 18:513–519, 1971.
154. Perry TL, Hansen S: Quantification of free amino compounds of rat brain: Identification of hypotaurine. J Neurochem 21:1009–1011, 1973.
155. Perry TL, Hansen S: Taurine content of epileptogenic foci in human brain. In Huxtable R, Barbeau A (eds): "Taurine." New York: Raven, 1976, pp 275–281.
156. Perry TL, Hansen S, Berry K, Mok D, Lesk D: Free amino acids and related compounds in biopsies of human brain. J Neurochem 18:521–528, 1971.
157. Perry TL, Hansen S, Kennedy J, Wada J, Thompson GB: Amino acids in human epileptogenic foci. Arch Neurol 32:752–754, 1975.
158. Peterson MB, Mead RJ, Welty JD: Free amino acids in congestive heart failure. J Mol Cell Cardiol 5:139–147, 1973.
159. Piha RS, Saukkonen H: The L-cysteine sulphinic acid decarboxylase activities in various areas of the brain. Suomen Kemistilehti B39:112–114, 1966.
160. Rassin DK, Gaull GE, Heinonen K, Raiha NCR: Milk protein quantity and quality in low birth weight infants. II. Effects on selected aliphatic amino acids in plasma and urine. Pediatrics 49:307–422, 1977.
161. Rassin DK, Sturman JA: Cystine sulfinic acid decarboxylase in rat brain: Effect of vitamin B_6 deficiency on soluble and particulate components. Life Sci 16:875–882, 1975.
162. Rassin DK, Sturman JA, Gaull GE: Taurine and other free amino acids in milk of man and other mammals. Early Human Development 2:1–13, 1978.
163. Read WO, Welty JD: Synthesis of taurine and isethionic acid by dog heart slices. J Biol Chem 237:1521–1522, 1962.
164. Remtulla MA, Applegarth DA, Clark DG, Williams IH: Analysis of isethionic acid in mammalian tissues. Life Sci 20:2029–2036, 1977.
165. Roches JC, Fassler A, Zumstein JR, Scollolog G, Hosli L: Action of taurine and glycine on strychnine-induced convulsions in rabbit. Experientia 34:902, 1978.
166. Roches JC, Zumstein HR, Fassler A, Scollo-Lavizzari G, Hosli L: Effects of taurine, glycine and GABA on convulsions produced by strychnine in the rabbit. Eur Neurol 18:26–32, 1979.
167. Sandberg PR, Willow M: Dose-dependent effects of taurine on convulsions induced by hypoxia in the rat. Neurosci Lett 16:297–300, 1980.
168. Sass NL, Martin WG: The synthesis of taurine from Sulfate III. Further evidence for the enzymatic pathway in the chick liver. Proc Soc Exp Biol Med 139:755–761, 1972.
169. Sbarbaro V: Effetti elettroclinici della taurina in alcuni pazienti epilepttici, risultati preliminari. Acta Neurol (Napoli) 29:33–37, 1974.
170. Scandurra R, Federici G, Dupre S, Cavallini D: Taurine and isethionic acid production in mammals. Bull Mol Biol Med 3:141–147, 1978.
171. Scandurra R, Politi L, Dupre S, Moriggi M, Barra D, Cavallini D: Comparative biological production of taurine from free cysteine and cysteine bound. Bull Mol Biol Med 2:172–177, 1977.
172. Schaffer SW, Sevem A, Chovan J: New chromatographic procedure for measuring isethionic acid levels in tissue. In Barbeau A, Huxtable R (eds): "Taurine and Neurological Disorders." New York: Raven, 1978, pp 25–28.
173. Schmidt SY, Berson EL, Hayes KC: Retinal degeneration in cats fed casein. 1. Taurine deficiency. Invest Ophthalmol 15:47–52, 1976.

174. Schousboe A, Fosmark H, Svenneby G: Taurine uptake in astrocytes cultured from dissociated mouse brain hemispheres. Brain Res 116:158–164, 1976.
175. Snyder SH: Brain peptides as neurotransmitters. Science 209:976–983, 1980.
176. Sörbo B, Heyman T: On the purification of cysteine sulphinic acid decarboxylase and its substrate specificity. Biochim Biophys Acta 23:624–627, 1957.
177. Striano S, Buscaino GA: Effects of taurine in human epilepsy, preliminary report. Acta Neurol (Napoli) 29:537–542, 1974.
178. Sturman JA: Taurine pool sizes in the rat: Effects of vitamin B-6 deficiency and high-taurine diet. J Nutr 103:1566–1580, 1973.
179. Sturman JA, Rassin DK, Gaull GE: Taurine in development. Life Sci 21:1–22, 1977a.
180. Sturman JA, Rassin DK, Gaull GE: Taurine in developing rat brain: Maternal-fetal transfer of (^{35}S) taurine and its fate in the neonate. J Neurochem 28:31–39, 1977b.
181. Sturman JA, Rassin DK, Gaull GE: Taurine in developing rat brain: Transfer of (^{35}S) taurine to pups via the milk. Pediat Res 11:28–33, 1977c.
182. Sumizu K: Oxidation of hypotaurine in rat liver. Biochem Biophys Acta 63:210–212, 1962.
183. Takahashi R, Nakane Y: Clinical trial of taurine in epilepsy. In Barbeau A, Huxtable R (eds): "Taurine and Neurological Disorders." New York: Raven, 1978, pp 375–386.
184. Thursby MH, Nevis AH: Anticonvulsant activity of taurine in electrically and osmotically induced seizures in mice and rats. Fed Proc 33:1494, 1974.
185. Thurston JH, Hauhart RE, Dirgo JA: Taurine: A role in osmotic regulation of mammalian brain and possible clinical significance. Life Sci 26:1561–1568, 1980.
186. Tsukada Y, Inoue N, Donaldson J, Barbeau A: Suppressive effects of various amino acids against ouabain-induced seizures in rats. Canad J Neurol Sci 1:214–221, 1974.
187. Urquhart N, Perry TL, Hansen S, Kennedy J: Passage of taurine into mammalian brain. J Neurochem 22:871–872, 1974.
188. Van Gelder NM: Antagonism by taurine of cobalt induced epilepsy in cat and mouse. Brain Res 47:157–165, 1972.
189. Van Gelder NM: Rectification of abnormal glutamate levels by taurine. In Huxtable R, Barbeau A (eds): "Taurine." New York: Raven, 1976, pp 293–302.
190. Van Gelder NM: Taurine, the compartmentalized metabolism of glutamic acid and the epilepsies. Canad J Physiol Pharmacol 56:362–374, 1978a.
191. Van Gelder NM: Glutamic acid and epilepsy: The action of taurine. In Barbeau A, Huxtable R (eds): "Taurine and Neurological Disorders." New York: Raven, 1978b, pp 387–402.
192. Van Gelder NM: Glutamic acid metabolism and epilepsy. Japan J NRA 6:(Supp) 163–176, 1980.
193. Van Gelder NM, Courtois A: Close correlation between changing content of specific amino acids in epileptogenic cortex of cats, and severity of epilepsy. Brain Res 43:477–484, 1972.
194. Van Gelder NM, Drujan BD: Alterations in the compartmentalized metabolism of glutamic acid with changed cerebral conditions. Brain Res 200:1–13, 1980.
195. Van Gelder NM, Koyama I, Jasper HH: Taurine treatment of spontaneous chronic epilepsy in a cat. Epilepsia 18:45–54, 1977.
196. Van Gelder NM, Sherwin AL, Sacks C, Andermann G: Biochemical observations following administration of taurine to patients with epilepsy. Brain Res 94:297–306, 1975.
197. Vessey DA: The biochemical basis for the conjugation of bile acids with either glycine and taurine. Biochem J 174:621–626, 1978.
198. Voaden MJ, Lake N, Marshall J, Morjaria B: Studies on the distribution of taurine and other neuroactive amino acids in the retina. Exp Eye Res 25:249–257, 1977.
199. Wada JA, Osawa T, Wake A, Corcoran ME: Effects of taurine on kindled amygdaloid seizures in rats, cats and photosensitive baboons. Epilepsia 16:229–234, 1975.
200. Wainer A: The production of sulfate from cysteine without the formation of free cysteine sulfinic acid. Biochem Biophys Res Comm 16:141–144, 1964.

201. Wainer A: The production of cysteine sulfinic acid from cysteine in vitro. Biochim Biophys Acta 104:405–412, 1965.
202. Wasson RG: Fly agaric and man. In Efron DH, Holmstedt B, Kline NS (eds): "Ethnopharmacologic Search For Psychoactive Drugs." New York: Raven, 1979, pp 405–414.
203. Welty JD, Read WO, Shaw EH: Isolation of 2-hydroxyethane sulfonic acid (isethionic acid) from dog heart. J Biol Chem 237:1160–1161, 1962.
204. Wheler GHT, Klein DC: Taurine release from the pineal gland is stimulated via a β-adrenergic mechanism. Brain Res 187:155–164, 1980.
205. Wheler GHT, Weller JL, Klein DC: Taurine: Stimulation of pineal N-acetyl transferase activity and melatonin production via a β-adrenergic mechanism. Brain Res 166:65–74, 1979.
206. Windholz M, Budavari S, Stroumtsos LY, Fertig MN: "The Merck Index: An Encyclopedia of Chemicals and Drugs." 9th Ed, New York: Merck, 1976.
207. Wu JY, Chen MS, Huang WM: Purification and immunochemical studies of cysteic acid decarboxylase and glutamate decarboxylase from bovine brain. Neurosci Abstr 4:454, 1978.
208. Wu JY, Moss LG, Chen MS: Tissue and regional distribution of cysteic and decarboxylase. Neurochem Res 4:201–212, 1979.
209. Yamaguchi K: Cysteine dioxygenase of rat liver: Some properties and its significance on cysteine metabolism of rat. In Cavallini D, Gaull GE, Zappia B (eds): "Natural Sulfur Compounds." New York: Plenum, 1980, pp 175–186.
210. Yamaguchi K, Sakakibara S, Asamizu J, Ueda I: Induction and activation of cysteine oxidase of rat liver: II. The measurement of cysteine metabolism in vivo and the activation of in vivo activity of cysteine oxidase. Biochim Biophys Acta 297:48–59, 1973.

The Role of Peptides and Amino Acids as Neurotransmitters, pages 99–116
©1981 Alan R. Liss, Inc., 150 Fifth Avenue, New York, NY 10011

Effects of TRH on Central Nervous System Function

George R. Breese, Robert A. Mueller, Richard B. Mailman, and Gerald D. Frye

An important aspect of processing of information in the central nervous system (CNS) is generally acknowledged to involve synaptic interactions of neurotransmitters. For many years, investigations in this area focused on the role of acetylcholine, monoamines, and amino acids as potential synaptic mediators. However, more recently another class of biological compounds, peptides, has also been suggested to play an essential role in brain activity. The initial evidence that such compounds were biologically important came from work demonstrating that peptides localized in the hypothalamus could influence the release of hormones from pituitary. The first of these factors isolated was thyrotropin-releasing hormone (TRH), which was identified as the tripeptide pyroglutamyl–histidyl–prolinamide [Burgus et al, 1969, 1970; Bohler et al, 1969]. Since these early investigations, more than a dozen other neuroactive peptides have been isolated from brain tissue and implicated in CNS function. Aspects of this rapidly expanding area of neurobiology are discussed in several recent reviews (eg [Prange et al, 1978; Snyder, 1980; Hökfelt et al, 1980]).

For several years, our laboratory has focused attention on the biological activities of TRH. When this compound was first isolated, TRH was believed to function primarily in the control of thyrotropin release from the pituitary [Bowers et al, 1971]. However, subsequent investigations which defined behavioral effects of TRH suggest that it also had extrahypothalamic actions [Plotnikoff et al, 1972; Breese et al, 1975; Winokur and Utiger, 1974]. This latter characterization provided the first evidence that releasing factors might have roles as neurotransmitters or neuromodulators distinct from their actions on the pituitary [Breese et al, 1974a; Prange et al, 1976].

Interactions of TRH With L-dihydroxyphenylalanine (L-dopa)

Early studies in our laboratory indicated that triiodothyronine could potentiate the behavioral effects of L-dopa in pargyline-pretreated mice [Breese et al, 1972]. When TRH was examined in this test, it too was found to enhance the actions of L-dopa (Fig. 1 [Plotnikoff et al, 1972, 1974a]). However, TRH was effective in potentiating these L-dopa-induced behaviors in hypophysectomized [Plotnikoff et al, 1972] as well as thyroidectomized animals [Plotnikoff et al, 1974b]. These results provided the first evidence of a behavioral action of TRH independent of the pituitary–thyroid axis. Other experiments demonstrated that tolerance to this action of TRH did not develop and that removal of many peripheral organs had no major effect on the TRH-induced increase in L-dopa activity [Plotnikoff et al, 1975].

One possible mechanism by which TRH might have enhanced the effects of L-dopa was by altering its distribution and metabolism. In order to examine this proposed mechanism, dopamine and L-dopa were measured in the brain of control and TRH-treated mice after L-dopa administration. L-dopa administration caused an increase in brain dopamine content, and TRH potentiated the change without increasing the amount of L-dopa in the brain [Breese et al, 1974a]. Since this increase in brain dopamine occurred concomitantly with peak behavioral activity, it may explain the TRH potentiation of L-dopa actions. However, the underlying neural mechanism responsible for this elevated content of dopamine induced by TRH has not yet been elucidated.

TRH Interactions With Centrally Acting Drugs

Subsequent studies explored interactions of TRH with a variety of other centrally acting drugs. Previous work had shown that chronic treatment with thyroid hormone enhanced the lethality of a number of psychotropic drugs [Breese et al, 1974a]. Nevertheless, neither acute nor chronic administration of TRH affected survival of animals treated with imipramine, chlorpromazine, pargyline, or phenelzine [Breese et al, 1974a]. These results suggested that TRH did not produce a hyperthyroid state, a view confirmed by direct measurement of thyroid function [Breese et al, 1974a].

In subsequent work, TRH was found to have profound effects on the actions of sedative–hypnotic drugs like pentobarbital and ethanol [Breese et al, 1974b; Prange et al, 1974, 1975; Breese et al, 1975]. For example, TRH antagonized pentobarbital-induced sleep, as illustrated by the return of righting reflex (Fig. 2 [Breese et al, 1975]). Similarly, TRH was found to reverse the hypothermia caused by pentobarbital [Breese et al, 1975]. One puzzling aspect of these data which has yet to be resolved is the lack of an apparent dose–response relationship for these actions of TRH (Fig. 1). In addition to its effects against pentobarbital-induced narcosis, TRH was subsequently shown to decrease the duration of ethanol-induced anesthesia and hypothermia and to reverse the ataxia induced by reserpine, chlorpromazine, and diazepam [Breese et al, 1975; Mailman et al, 1980]. Further, pharmacokinetic studies indicated that the action of TRH to antagonize pentobarbital and ethanol was not due to an alteration in the metabolism or distribution of these depressants [Breese et al, 1975; Cott et al, 1976;

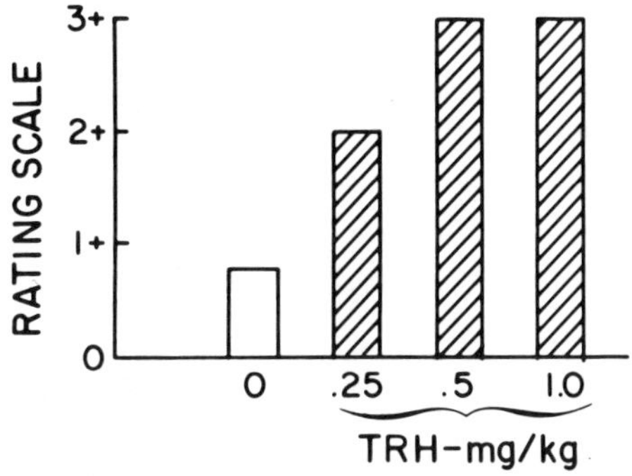

Fig. 1. Effect of TRH on the dopa-potentiation test. Mice were treated with pargyline (40 mg/kg) 1 hr before receiving various doses of TRH (lined bars). All animals received dihydroxyphenylalanine (L-dopa, 100 mg/kg) 1 hr after TRH or saline. Each bar represents from two to four groups of four mice each. (from [Breese et al, 1974a])

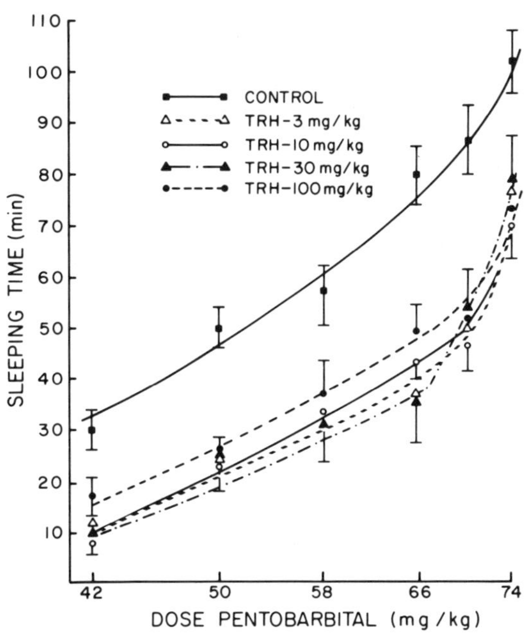

Fig. 2. Effect of various doses of TRH on the dose–response relationships of pentobarbital sleeping time. Doses of TRH were administered 5 min before injection of pentobarbital. Each point represents mean ± SEM sleeping time of 10 to 40 mice. All mean values for TRH are significantly different from control ($P < 0.001$). (from [Breese et al, 1975])

Frye et al, 1981]. Moreover, the analeptic actions of TRH were not attenuated by hypophysectomy, or emulated by thyroid hormone treatment (Table I). These experiments provided additional support for the contention that TRH had effects on the CNS distinct from its neuroendocrine actions [Breese et al, 1975; Cott et al, 1976].

Localization of TRH and Its Receptors in Brain

Localization of TRH in brain areas other than hypothalamus [Winokur and Utiger, 1974; Oliver et al, 1974; Jackson and Reichlin, 1974; Hökfelt et al, 1975; Mailman et al, 1980] provided further support for extrahypothalamic functions of TRH in brain (Table II). From these data, approximately 70% of the total TRH in brain was demonstrated to be present in the areas of brain other than hypothalamus [Winokur and Utiger, 1974]. While there has been controversy concerning the authenticity of immunoreactive TRH in these extrahypothalamic brain areas [Youngblood et al, 1978; Emerson et al, 1978], the weight of recent evidence seems to indicate the presence of authentic TRH in extrahypothalamic areas of the CNS [Kellokumpu et al, 1980].

Another important line of investigation which was supportive of an extrahypothalamic role for TRH was the mapping of TRH-binding sites in the brain (Table II [Burt and Snyder, 1975; Burt and Taylor, 1980]). These radioligand studies showed that TRH binding was present in several areas of the CNS other than hypothalamus and that the binding was similar to that observed in the pituitary. The latter finding is consistent with our earlier data demonstrating that several chemical analogs of TRH which release TSH (presumably by acting on TRH receptors in the pituitary) are also those effective in reversing pharmacological actions of pentobarbital or ethanol [Breese et al, 1975; Cott et al, 1976]. Recently, a functional differentiation of TRH receptor sites was suggested from

TABLE I. Effect of Hypophysectomy on the Actions of TRH to Antagonize Pentobarbital- and Ethanol-Induced Sleep and Hypothermia

Treatment[a]	Sleeping time (min ± SEM)	Rectal temperature (°C ± SEM)
Saline + pentobarbital	68.0 ± 8.1	32.3 ± 0.4
TRH + pentobarbital	35.7 ± 2.2*	34.8 ± 0.3*
Saline + ethanol	58.8 ± 9.6	32.9 ± 0.4
TRH + ethanol	31.2 ± 2.7*	34.0 ± 0.3*

[a]All mice were hypophysectomized 4–5 days before receiving 45 mg/kg of pentobarbital or 4.3 g/kg of ethanol.

*$P < 0.05$ when compared to hypophysectomized mice that received saline.

TABLE II. Regional Distribution of TRH Content and Binding in Mammalian Brain*

Region	TRH binding (fmol bound/g)	TRH content (ng/mg protein)
Nucleus accumbens	313 ± 61	—
Caudate	141 ± 29	0.05 ± 0.007
Septal region	138 ± 46	—
Olfactory tubercle	89 ± 18	—
Cerebral cortex	29 ± 18	0.02 ± 0.004
Hippocampus	46 ± 14	0.04 ± 0.01
Amygdala	55 ± 16	0.11 ± 0.06
Medulla	45 ± 4	0.01 ± 0.002
Ventral hypothalamus	20 ± 10	2.15 ± 0.42
Cerebellum	21 ± 9	0.01 ± 0.001

*Modified from Burt and Taylor [1980] and Kubek et al [1977]. Binding determinations were in sheep brain and content measures were in human brain.

TABLE III. Effect of Central Administration of Atropine Methyl Nitrate (AMN) on the Analeptic Effects of TRH

		Pretreatment (min ± SEM)	
Treatment[a]	TRH	Saline	AMN
Pentobarbital	No	45 ± 5.3	45 ± 8.8
Pentobarbital	Yes	17 ± 2.2*	45 ± 6.4**
Ethanol	No	109 ± 8.0	110 ± 6.9
Ethanol	Yes	69 ± 4.3*	117 ± 5.4**

[a]The dose of pentobarbital sodium was 55 mg/kg and that for ethanol was 4.6 g/kg. TRH (10 mg/kg) was given 10 min after the drugs. Atropine methyl nitrate (AMN; 20 μg to mice given pentobarbital and 10 μg to mice given ethanol) was administered intracisternally 10 min before the depressants. There are at least 15 animals in each group. (Modified from [Breese et al, 1975 and Cott et al, 1976])
*$P < 0.001$ when compared to animals not receiving TRH.
**$P < 0.001$ when compared to saline.

work showing that administration of TRH into specific extrahypothalamic areas of the CNS resulted in antagonism of pentobarbital-induced narcosis, whereas injection into other sites did not support this action of TRH [Kalivas and Horita, 1979]. Injection of TRH into the septum was found to be most effective against pentobarbital-induced sleep, while other extrahypothalamic sites were responsible for the TRH-induced antagonism of the hypothermia [Kalivas and Horita,

1980]. Furthermore, Nagai et al [1980] have reported that TRH administration causes a significant decrease of glucose utilization in several extrahypothalamic areas of the brain. Interestingly, TRH actually increased glucose utilization in the brain of rats treated with pentobarbital. These latter data are further evidence for the diverse and potent actions of TRH.

One might assume that those extrahypothalamic areas having high concentrations of TRH receptors would also have the greatest concentrations of endogenous TRH. However, it is apparent when comparing literature values of the distribution of TRH in the brain with the density of TRH-binding sites that this contention does not necessarily hold (Table II). Therefore, the literature would benefit considerably from a thorough examination of the correlation of these two parameters.

Entry of TRH Into the Brain after Peripheral Administration

Several early investigations demonstrated that small doses of TRH administered directly into the brain were as effective as much larger doses given peripherally [Breese et al, 1975; Cott et al, 1976]. Although these latter findings were consistent with TRH's having central actions, it remained to be shown if the behavioral effects resulting from parenteral administration of TRH were actually due to TRH's entering into the brain. Initial studies were unable to detect the accumulation of labeled TRH in the brain after administration of low doses of [^{3}H]TRH into the internal carotid artery [Steiner et al, 1974; Cornford et al, 1978]. Our laboratory also attempted to detect labeled TRH in the brain after peripheral administration of pharmacological doses, but these efforts were also unsuccessful [unpublished data]. A possible reason for this failure was that the specific activity of TRH was insufficient to assess the passage of small quantities of TRH into the brain. Regardless, we have recently been able to demonstrate the presence of increased amounts of TRH immunoreactivity in the brain after peripheral administration of pharmacological doses of TRH, utilizing a TRH radioimmunoassay of brain tissue extracts (Fig. 3 [Mailman et al, 1979, 1980]). It is not presently known whether this is authentic TRH or a metabolite, but these data clearly indicate that TRH enters the brain after peripheral administration.

Physiological Alterations Induced by TRH

Several types of changes in motor function have been described after TRH administration [Breese et al, 1975; Wei et al, 1975b; Segal and Mandell, 1974; Miyamoto and Nagawa, 1977]. Forepaw tremor, head and body shaking (wet dog shakes), changes in tail position, and increased activity are all reported to be increased by TRH. Cooper and Boyer [1978] recently showed that TRH facilitated spinal reflexes, raising the possibility that an action of TRH at this level of the CNS could contribute to some aspects of the behavioral syndrome induced by TRH.

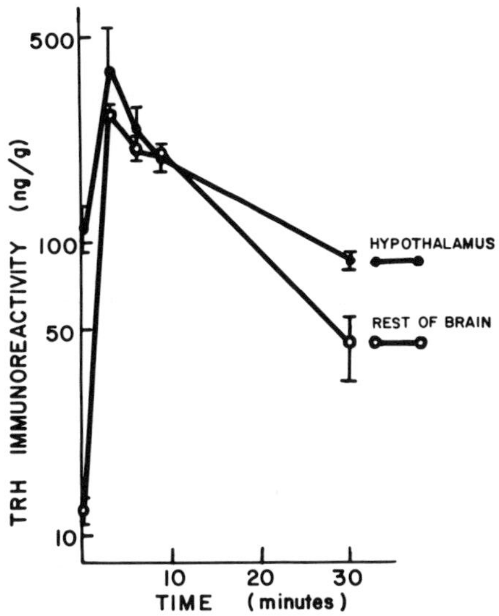

Fig. 3. Levels of immunoreactive TRH in brain before and after intravenous administration of synthetic TRH (20 mg/kg). TRH content was significantly elevated in rest of brain at all time points and in the hypothalamus through 10 min after injection ($P < 0.05$).

Investigators have suggested that at least some of the motor changes may be associated with stimulation of pathways important to temperature control [Wei et al, 1975b]. Early studies demonstrated that TRH antagonized the hypothermia induced by pentobarbital and other centrally acting drugs [Breese et al, 1975]. While peripheral TRH administration to conscious animals seldom causes hyperthermia, such a response is observed when TRH is administered to mice treated with reserpine [unpublished data]. In contrast, Carino et al [1976] observed hyperthermia with microinjections of TRH into the hypothalamus of rats. Based on the similarity of the changes in motor function and effects on body temperature, Wei et al [1975a] suggested that TRH might be involved in the morphine abstinence syndrome. Regardless of the interpretation, recent results suggest a close anatomical relationship between areas responsible for the shaking response and the temperature alterations induced by TRH [Kalivas and Horita, 1980].

In addition to its effects on temperature and motor function, TRH has been shown to alter several other physiological parameters. One of these is respiration [Kraemer et al, 1976; Horita et al, 1976]. This prominent action appears to be associated with mesencephalic reticular areas, but specific sites for increases in respiration have not been elucidated. Metcalf and Myers [1976] suggested that the TRH-induced tachypnea observed in cats might explain the hypothermia noted in this species [Metcalf, 1974]. Changes in blood pressure have also been demonstrated after central administration of TRH [Beale et al, 1977], and TRH has been shown to induce anorexia in food-deprived rats (Fig. 4 [Vogel et al, 1979]). This latter behavioral change may be linked to the report by Smith et al [1977] that TRH increases gut motility.

Effects of TRH on Operant Responding

Early studies suggested that TRH did not alter responding in various operant tasks when testing was done 60 min after injection [Breese et al, 1974a]. However, subsequent studies indicated that the time between TRH administration and testing was a critical factor as to whether TRH induces a change in operant performance [Vogel et al, 1979]. For example, immediately after injection, TRH produces a marked reduction in food-reinforced operant responding, but rates subsequently return to normal within 60 min (Fig. 5). Since TRH causes major alterations in motor function, such changes were considered responsible for the reduced performance in this task. However, evidence seemed to discount this possibility. For example, self-stimulation responding was unaffected by TRH, using response rates that were similar to those used in the food-motivated paradigm. Furthermore, avoidance responding in a shuttle box was unaffected by TRH [Vogel et al, 1979]. However, subsequent work demonstrated that TRH

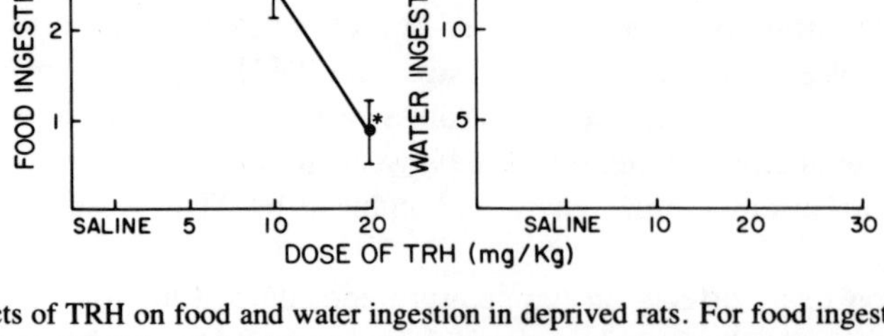

Fig. 4. Effects of TRH on food and water ingestion in deprived rats. For food ingestion, rats were placed on a 20 hr schedule of food deprivation and the amount of food consumed during 30 min after introduction of food determined. Water consumption was determined in rats placed on a 22 hr schedule of fluid deprivation. * $P < 0.01$ when compared to saline response. (from [Vogel et al, 1979])

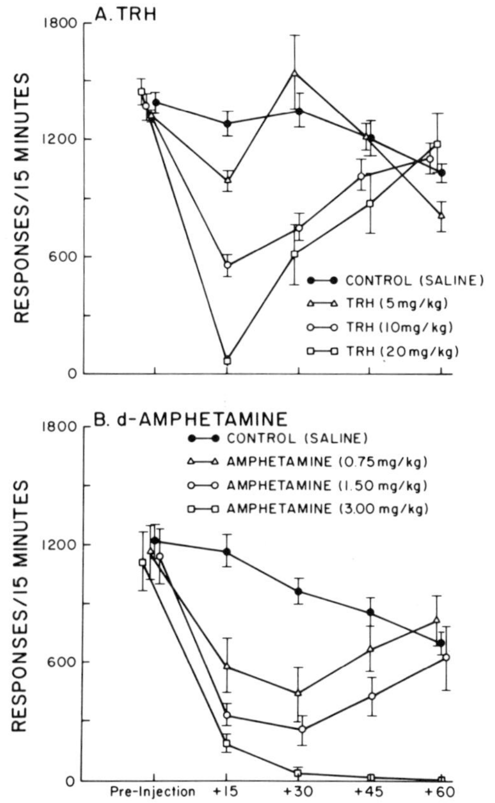

Fig. 5. Effects of TRH (A) and d-amphetamine (B) on bar press responses maintained by FR-30 schedule of food reinforcement. Responding during the first 15 min after the 10 or 20 mg/kg doses of TRH was significantly different from control period on the previous day of testing ($P < 0.01$). All doses of d-amphetamine produced a significant reduction in responding ($P < 0.01$). Drug doses are administered ip (from [Vogel et al, 1979]).

induced an anorexia (Fig. 4) that paralleled the deficit produced in the food-reinforced task. Thus, this reduced food intake is believed to be responsible for the TRH-induced change in operant performance [Vogel et al, 1979].

Several observations indicated similarities in the actions of TRH and d-amphetamine. For example, both inhibit operant responding (Fig. 5), increase locomotion, and cause anorexia [Hollister et al, 1974, 1975]. For this reason, several studies were undertaken to compare the actions of these two compounds in more detail. Despite the superficial similarities shown to exist, other data clearly differentiated their mechanisms of action. For example, 6-hydroxydopamine treatment was found to antagonize d-amphetamine-induced anorexia and increased locomotor activity, but it did not reduce the effects of TRH on these measures [Vogel et al, 1979]. Furthermore, at low rates of self-stimulation, d-amphetamine increased responding, whereas TRH had no effect or decreased the rate [Vogel et al, 1979].

Since TRH had been demonstrated to reduce the locomotor depressant actions of benzodiazepines [Breese et al, 1975], studies were undertaken to determine if TRH might also attenuate the change in responding induced by this drug class in an acute lick-suppression punishment task. Unexpectedly, TRH actually enhanced the action of certain doses of chlordiazepoxide in this model. Since this preliminary evidence suggested that TRH, itself, might possess antipunishment activity, its actions in this conflict task were more thoroughly characterized [Vogel et al, 1980]. It was demonstrated that doses greater than 10 mg/kg of TRH significantly increased punished drinking, with maximal responses of TRH occurring at 40 or 80 mg/kg (Fig. 6). Examination of the time course of this response indicated that the activity of TRH was apparent within 15 min and had a detectable effect that persisted through 60 min. By 2 hr, values had returned

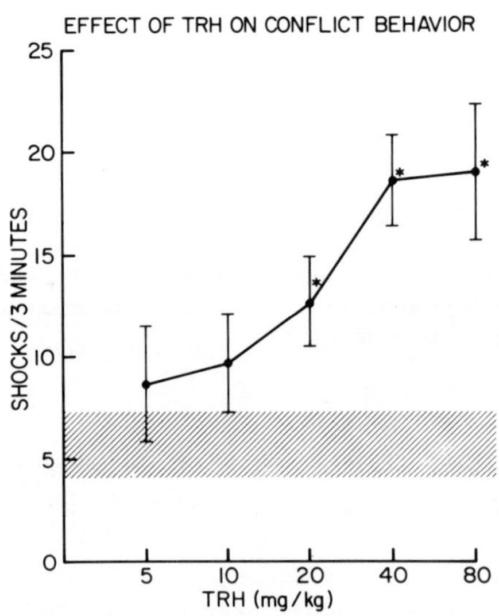

Fig. 6. Dose–response relationship of TRH on punished behavior. Doses of TRH or saline (shaded area) were administered ip 30 min before testing. Values represent mean ± SEM of at least eight rats. *$P < 0.05$ when compared to saline. (from [Vogel et al, 1980]).

to control. Since neither thyrotropin nor triiodothyronine was active in this paradigm, this action of TRH did not appear to be mediated by the pituitary–thyroid axis. Further, doses of TRH from 1 to 25 μg (administered intracisternally) also had potent effects on punished responding, providing evidence that this action is centrally mediated. Because this conflict paradigm is sensitive to clinically useful anxiolytic agents [Vogel et al, 1971], it is possible that this effect of TRH is related to the "relaxation" reported in humans after intravenous administration [Wilson et al, 1973].

TRH Interactions with Neurotransmitter Systems

Evidence from a variety of neurobiological studies has suggested that the effects of TRH are mediated by interactions with several neurotransmitter systems in brain. The earliest studies demonstrated that TRH increased norepinephrine turnover without affecting amine content [Horst and Spirt, 1974; Breese et al, 1974a, 1975; Keller et al, 1974]. However, efforts to block the analeptic actions of TRH with adrenergic receptor blockers were not effective [Breese et al, 1975; Cott et al, 1976]. Similarly, α- and β- adrenergic receptor agonists did not reverse the anesthesia induced by barbiturates and ethanol, as did TRH. Furthermore, as has already been mentioned, destruction of adrenergic fibers with 6-hydroxydopamine treatment did not alter the locomotor activity or anorexia induced by TRH [Vogel et al, 1979]. Thus, the observed biochemical change in noradrenergic mechanisms caused by TRH has not yet been related to a specific physiological or behavioral action.

Cott and Engel [1977] showed that drugs which mimic γ-aminobutyric acid (GABA) would antagonize the analeptic actions of TRH to reduce ethanol-induced sleep and hypothermia. From the results, they proposed that the analeptic actions of TRH may be related to an inhibition of GABA function. In agreement with these earlier studies, Yarbrough [1979] confirmed that baclophen reduced TRH-induced tremor, antagonized the ability of TRH to decrease the sleep induced by depressants, and prevented the elevated respiration caused by TRH in pentobarbital-treated rats. Interestingly, other GABA-mimetics, such as muscimol, did not alter these actions of TRH [Yarbrough, 1979]. Since there are questions concerning the mechanism of action of baclophen, it is uncertain what the exact nature of TRH-GABA interactions may be.

TRH is also believed to alter serotonergic receptor function, since it will potentiate the actions of 5-hydroxytryptophan and other serotonergic agonists. Recently, our laboratory has found additional evidence to support an important relationship between TRH and serotonergic functions. Administration of 5,7-dihydroxytryptamine (5,7-DHT) into the brain of neonatal rats will cause marked decreases in serotonin concentration in brain [Breese et al, 1978]. However, this treatment also decreases the TRH immunoreactivity in extrahypothalamic areas of brain, but not in the hypothalamus (Fig. 7). This change persists through at

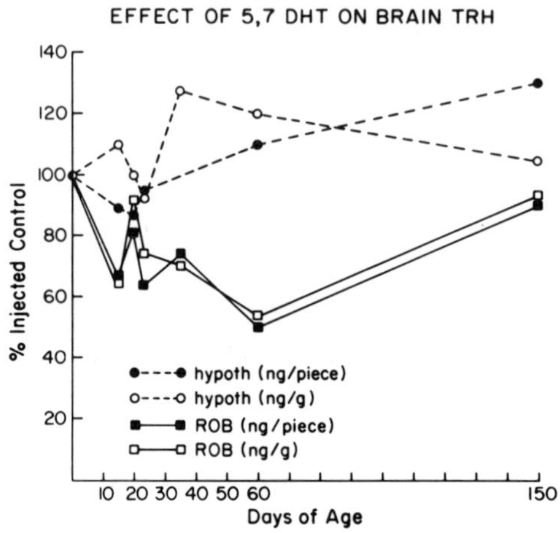

Fig. 7 Effect of neonatal treatment with 5,7-dihydroxytryptamine on TRH content in hypothalamus (HYPOTH) and rest of brain (ROB). Rat pups were treated with 5,7-DHT (50 μg) in the presence of pargyline (50 mg/kg) when 3 days of age (see [Breese et al, 1978]). Content in the hypothalamus was not significantly affected. TRH content in the ROB was significantly reduced through 60 days of age but not at 150 days of age. [unpublished data].

least 60 days of age. However, it remains to be documented that the 5,7-DHT-induced decreases in immunoreactive TRH in the neonatal rat is a change in TRH itself, rather than in another immunoreactive substance. Nevertheless, this finding could be of considerable importance, especially since Hökfelt et al [1980] have reported that there is TRH immunoreactivity localized in serotonergic fibers.

Finally, considerable data have linked the effects of TRH to central cholinergic mechanisms. For example, the analeptic action of TRH has been shown to be antagonized by anticholinergic drugs (Table III; [Breese et al, 1975; Cott et al, 1976; Horita et al, 1976]). Furthermore, muscarinic-blocking drugs have been reported to abolish the actions of TRH to activate the EEG in rabbits [Beale et al, 1977] and to block the TRH-induced increase in glucose utilization in pentobarbital-treated rats [Nagai et al, 1980]. Thus, these findings suggest that the analeptic actions of TRH may be mediated through a cholinergic mechanism. In accord with this view, Schmidt [1977] has shown that TRH reverses the decreased cholinergic activity caused by depressant drugs. Further, TRH is reported to reduce acetylcholine content and increase turnover in cortical regions [Malthe-Sorenssen et al, 1978] and to enhance the excitatory actions of acetylcholine on individual neurons [Yarbrough and Singh, 1978]. All these data are suggestive of a modulatory role for TRH on synaptic processes involving cholinergic neurons.

Another observation that was initially believed to be associated with an effect of TRH on cholinergic fibers was the in vivo increase of cerebellar guanosine 3′,5′-monophosphate (cGMP) concentration following TRH administration [Mailman et al, 1979]. Like many other effects of TRH, this change was independent of pituitary function and was also induced by active analogs of TRH. However, unlike the analeptic action of TRH, the increase in cGMP was not antagonized by atropine. Furthermore, it was subsequently shown that these cerebellar cGMP perturbations were largely secondary to the increased motor activity induced by TRH (Fig. 8; [Lundberg et al, 1979; Mailman et al, 1980]). Whereas paralysis eliminated the TRH-induced increase in cerebellar cGMP, this treatment revealed an increase in cGMP in the brain stem that was not otherwise detectable (Fig. 9). Thus, these series of experiments emphasize the necessity of controlling afferent stimuli when assessing drug effects on brain cyclic nucleotides. These data also suggest that complex interactions induced by TRH and other neural pathways can influence the magnitude of the neural response [Breese et al, 1979]. Additional studies are underway to determine if these findings reflect a general biological phenomenon by which one can index transmission of neural messages.

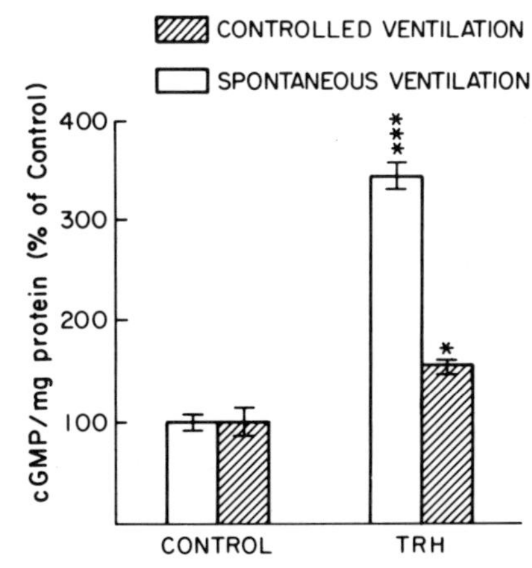

Fig. 8. Effect of TRH on cGMP content in cerebellum of paralyzed (controlled ventilation) and spontaneously active rats (spontaneous ventilation). After paralysis with d-tubocurarine, rats were ventilated to maintain normal arterial pH, CO_2, and O_2 tensions. Rats were killed with microwave fixation. TRH treated rats received 10 mg/kg, intravenously. Control spontaneously ventilated rats had cerebellar cGMP values of 5.6 ± 0.4 pmoles/mg of protein; controlled ventilated rats had 1.75 ± 0.24 pmoles/mg of protein. $^*P < 0.05$; $^{***}P < 0.001$ relative to corresponding control group. (modified from [Lundberg et al, 1979])

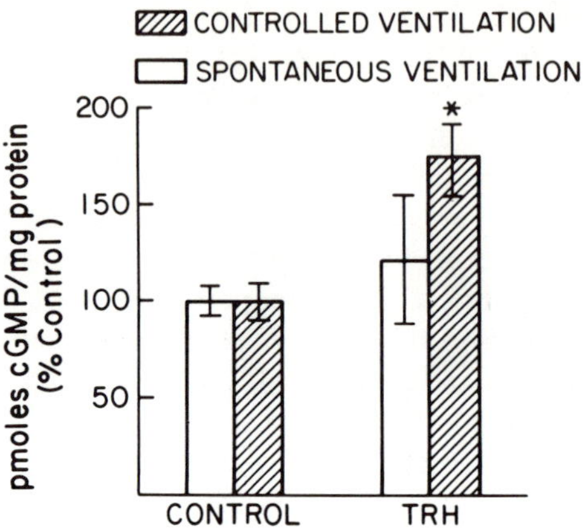

Fig. 9. Effect of TRH on cGMP content in brain stem of spontaneously active (spontaneous ventilation) and paralyzed (controlled ventilation) rats. Procedures are described in Figure 8. Rats received 10 mg/kg of TRH or saline intravenously. For the brain stem, control spontaneously ventilated rats had cGMP values of 0.81 ± 0.09 pmoles/mg of protein; control ventilated rats had values of 0.53 ± 0.05 pmoles/mg of protein [unpublished data]. *$P < 0.05$ relative to corresponding control group.

SUMMARY

Evidence has been reviewed which strongly suggests that TRH functions as a neurotransmitter or neuromodulator in the mammalian central nervous system. Both the peptide and its receptor sites are located in the brain. Furthermore, it has potent actions to modify the effects of many neuropharmacological agents and can itself cause alterations in functions mediated by the CNS. Data clearly indicate that many of these actions of TRH are not dependent on the pituitary–thyroid axis. Various studies of neurotransmitter interactions with TRH have provided evidence that noradrenergic, serotonergic, GABAergic, and cholinergic systems may be influenced by TRH or mediate some of its actions. More than likely, other transmitters will be implicated as the complex actions of TRH are more thoroughly investigated and understood. One set of experiments suggested that TRH may be localized in serotonergic fibers. Such findings provide strong support for the view that TRH has a role in the physiology of the CNS.

In spite of the progress that has been made, there are several questions to be answered about mechanisms of synthesis, storage, release, and inactivation of TRH. Furthermore, physiological studies would be greatly facilitated if a specific antagonist of TRH actions were available. Controversy still exists about active forms of TRH in situ and the methods by which TRH can be measured in tissue. Future investigations which resolve these difficulties and questions should facilitate our understanding of the role of TRH in brain function and its complex interactions with other neural mechanisms.

ACKNOWLEDGMENTS

Unpublished data presented were supported by USPHS Grants HD-03110, MH-23216, AA-02334 and the North Carolina Alcoholism Research Authority (No. 7806 and No. 7915).

REFERENCES

1. Beale JS, White RP, Huang S: EEG and blood pressure effects of TRH in rabbits. Neuropharmacology 16:449–506, 1977.
2. Bohler J, Enzmann F, Folkers K, Bowers CY, Schally AV: The identity of chemical and hormonal properties of the thyrotropin releasing hormone and pyroglutamyl-histidyl-prolineamide. Biochem Biophys Res Commun 37:705, 1969.
3. Bowers CY, Friesen HG, Hwang P, Guyda HJ, Folkes K: Prolactin and thyrotropin release in man by synthetic pyroglutamyl-histidyl-prolinamide. Biochem Biophys Res Comm 45:1033–1041, 1971.
4. Breese GR, Cooper BR, Prange AJ, Cott JM, Lipton MA: Interactions of thyrotropin releasing hormone with centrally acting drugs. In Prange AJ (ed): "The Thyroid Axis, Drugs, and Behavior." New York: Raven Press, 1974a. pp 115–127.
5. Breese GR, Cott JM, Cooper BR, Prange AJ, Lipton MA: Antagonism of ethanol narcosis by thyrotropin-releasing hormone. Life Sci 14:1053–1063, 1974b.
6. Breese GR, Cott JM, Cooper BR, Prange AJ Jr, Lipton MA, Plotnikoff NP: Effects of thyrotropin-releasing hormone (TRH) on the actions of pentobarbital and other centrally acting drugs. J Pharmacol Exp Ther 193:11–22, 1975.
7. Breese GR, Mueller RA, Mailman RB: Effect of dopaminergic agonists on *in vivo* cyclic nucleotide content: Relation of guanosine 3′:5′-monophosphate (cGMP) changes in cerebellum to behavior. J Pharmacol Exp Ther 209:262–270, 1979.
8. Breese GR, Vogel RA, Mueller RA: Biochemical and behavior alterations in developing rats treated with 5,7-dihydroxytryptamine. J Pharmacol Exp Ther 205:587–595, 1978.
9. Breese GR, Traylor TD, Prange AJ: The effect of triiodothyronine on the disposition and actions of imipramine. Psychopharmacologia 25:101–111, 1972.
10. Burgus R, Dunn TF, Desiderio D, Guillemin R: Structure moleculaire du facteur hypothalamique hypophysiotrope TRP d'origine ovine: Mise en evidence par spectrometrie de masse de la sequence PCA-His-Pro-NH_2. CR Acad Sci (Paris) 269:1870–1873, 1969.
11. Burgus R, Dunn TF, Desiderio D, Ward DN, Vale W, Guillemin R: Characterization of ovine, hypothalamic hypophysiotropic TSH releasing factor. Nature (London) 226:321–325, 1970.
12. Burt DR, Snyder SH: Thyrotropin-releasing hormone (TRH): Apparent receptor binding in rat brain membranes. Brain Res 93:309, 1975.
13. Burt DR, Taylor RL: Binding sites for thyrotropin-releasing hormone in sheep nucleus accumbens resemble pituitary receptors. Endocrinology 106:1416–1423, 1980.
14. Carino MA, Smith JR, Weick BG, Horita A: Effects of thyrotropin-releasing hormone microinjected into various brain areas of conscious and pentobarbital pretreated rabbits. Life Sci 19:1687–1692, 1976.
15. Cooper BR, Boyer CE: Stimulant action of thyrotropin releasing hormone on cat spinal cord. Neuropharmacology 17:153–156, 1978.

16. Cornford EM, Braun LD, Crane PD, Oldendorf WH: Blood–brain barrier restriction of peptides and the low uptake of enkephalins. Endocrinology 103:1297–1303, 1978.
17. Cott JM, Breese GR, Cooper BR, Barlow TS, Prange AJ Jr: Investigations into the mechanism of reduction of ethanol sleep by thyrotropin-releasing hormone (TRH). J Pharmacol Exp Ther 196:594–604, 1976.
18. Cott J, Engel J: Antagonism of the analeptic activity of thyrotropin releasing hormone by agents which enhance GABA transmission. Psychopharmacology 52:145–149, 1977.
19. Emerson CH, Frohman LA, Szabo M, Thakkar I: TRH immunoreactivity in human urine: Evidence for dissociation from TRH. J Clin Endocrinol Metab 45:392–399, 1977.
20. Frye GD, Luttinger D, Nemeroff C, Vogel R, Prange A, Breese G: Modification of the actions of ethanol by centrally active peptides. Peptides (in press), 1981.
21. Hökfelt T, Fuxe K, Johansson O, Jeffcoate S, White N: Thyrotropin releasing hormone (TRH)-containing nerve terminals in certain brain stem nuclei and in the spinal cord. Neurosci Lett 1:133–139, 1975.
22. Hökfelt T, Johansson O, Ljungdahl Å, Lundberg JM, Schultzberg M: Peptidergic neurones. Nature (London) 284:515–521, 1980.
23. Hollister AS, Breese GR, Cooper BR: Comparison of tyrosine hydroxylase and dopamine-β-hydroxylase inhibition with the effects of various 6-hydroxydopamine treatments on d-amphetamine induced locomotor activity. Psychopharmacologia 36:1–16, 1974.
24. Hollister AS, Ervin GN, Cooper BR, Breese GR: The roles of monoamine neural systems in the anorexia induced by d-amphetamine and related compounds. Neuropharmacology 14:715–723, 1973.
25. Horita A, Carino MA, Chesnut RM: Influence of thyrotropin-releasing hormone (TRH) on drug-induced narcosis and hypothermia in rabbits. Psychopharmacology 49:57–62, 1976.
26. Horst WD, Spirt N: A possible mechanism for the antidepressant activity of thyrotropin releasing hormone. Life Sci 15:1073–1082, 1974.
27. Jackson IMD, Reichlin S: Thyrotropin-releasing hormone (TRH): Distribution in hypothalamic and extrahypothalamic brain tissues of mammalian and submammalian chordates. Endocrinology 95:854–862, 1974.
28. Kalivas PW, Horita A: Thyrotropin-releasing hormone: Central site of action of pentobarbital narcosis. Nature (London) 278:461–463, 1979.
29. Kalivas PW, Horita A: Thyrotropin-releasing hormone: Neurogenesis of actions in the pentobarbital narcotized rat. J Pharmacol Exp Ther 212:203–210, 1980.
30. Keller H, Bartholini G, Pletscher A: Enhancement of cerebral noradrenaline turnover by thyrotropin-releasing hormone. Nature (London) 248:528–529, 1974.
31. Kellokumpu S, Vuolteenaho O, Leppaluoto J: Behavior of rat hypothalamic and extrahypothalamic immunoreactive TRF in thin-layer chromatography and high pressure liquid chromatography (HPLC). Life Sci 26:475–480, 1980.
32. Kraemer GW, Mueller RA, Breese GR, Prange AJ, Lewis JK, Morrison H, McKinney WT: Thyrotropin releasing hormone: Antagonism of pentobarbital narcosis in the monkey. Pharmacol Biochem Behav 4:709–712, 1976.
33. Kubek MJ, Lorincz MA, Wilber JF: The identification of thyrotropin-releasing hormone (TRH) in hypothalamic and extrahypothalamic loci of the human nervous system. Brain Res 126:196–200, 1977.
34. Lundberg D, Breese GR, Mailman RB, Frye GD, Mueller RA: Depression of some drug-induced *in vivo* changes in cerebellar guanosine-3′,5′-monophosphate (cGMP) by control of motor and respiratory changes. Mol Pharmacol 15:246–256, 1979.
35. Mailman RB, Frye GD, Mueller RA, Breese GR: Changes in brain guanosine-3′,5′-monophosphate (cGMP) content by thyrotropin-releasing hormone. J Pharmacol Exp Ther 208:169–175, 1979.
36. Mailman RB, Frye GD, Mueller RA, Breese GR: The effects of thyrotropin-releasing hormone

(TRH) and other drugs on the actions of alcohol. In Begleiter H (ed): "Biological Effects of Alcohol." New York: Plenum, 1980, pp 509–522.

37. Malthe-Sorenssen D, Wood PL, Cheney DL, Costa E: Modulation of the turnover rate of acetylcholine in rat brain by intraventricular injections of thyrotropin releasing hormone, somatostatin, enurotensin and angiotensin II. J Neurochem 31:685–691, 1978.
38. Metcalf G: TRH: A possible mediator of thermoregulation. Nature (London) 252:310–311, 1974.
39. Metcalf G, Myers RD: A comparison between the hypothermia induced by intraventricular injections of thyrotropin-releasing hormone, noradrenaline or calcium ions in unanesthetized cats. Br J Pharmacol 58:489–495, 1976.
40. Miyamoto M, Nagawa Y: Mesolimbic involvement in the locomotor stimulant action of thyrotropin-releasing hormone (TRH) in rats. Eur J Pharmacol 44:143–152, 1977.
41. Nagai Y, Narumi S, Nagawa Y, Sakurada O, Ueno H, Ishii S: Effect of thyrotropin-releasing hormone (TRH) on local cerebral glucose utilization by the autoradiographic 2-deoxy[^{14}C]glucose method, in conscious and pentobarbitalized rats. J Neurochem 35:963–971, 1980.
42. Oliver C, Eskay RL, Ben-Jonathan N, Porter JC: Distribution and concentration of TRH in the rat brain. Endocrinology 96:540–546, 1975.
43. Plotnikoff NP, Breese GR, Prange AJ Jr: Thyrotropin releasing hormone (TRH): DOPA potentiation and biogenic amine studies. Pharmacol Biochem Behav 3:665–670, 1975.
44. Plotnikoff NP, Prange AJ, Breese GR, Anderson MS: Enhancement of DOPA activity by a hypothalamic hormone, thyrotropin releasing hormone, TRH. Science 178:417–418, 1972.
45. Plotnikoff NP, Prange AJ, Breese GR, Anderson MS, Wilson IC: The effects of TRH on DOPA response in normal, hypophysectomized, and thyroidectomized animals. In Prange AJ (ed): "The Thyroid Axis, Drugs, and Behavior." New York: Raven Press, 1974a, pp 103–113.
46. Plotnikoff NP, Prange AJ, Breese GR, Wilson IC: Thyrotropin releasing hormone: Enhancement of DOPA activity in thyroidectomized rats. Life Sci 14:1271–1278, 1974b.
47. Prange AJ Jr, Breese GR, Cott JM, Martin BR, Cooper BR, Wilson IC, Plotnikoff NP: Thyrotropin-releasing hormone: Antagonism of pentobarbital in rodents. Life Sci 14:447–455, 1974.
48. Prange AJ Jr, Breese GR, Jahnke GD, Martin BR, Cooper BR, Cott JM, Wilson IC, Alltop LB, Lipton MA, Bissette G, Nemeroff CB, Loosen PT: Modification of pentobarbital effects by natural and synthetic polypeptides: Dissociation of brain and pituitary effects. Life Sci 16:1907–1914, 1975.
49. Prange AJ, Breese GR, Wilson IC, Lipson MA: Brain behavioral effects of hypothalamic releasing hormones: A generic hypothesis. In Stumpf WE and Grant LD (eds): "Anatomical Neuroendocrinology." Basel: Karger, 1976, pp 357–366.
50. Prange AJ, Nemeroff CB, Lipson MA, Breese GR, Wilson IC: Peptides and the central nervous system. In Iverson LL et al (eds): "Handbook of Psychopharmacology." New York: Plenum, 1978, pp 1–107.
51. Schmidt DE: Effects of thyrotropin-releasing hormone (TRH) on pentobarbital induced decrease in cholinergic neuronal activity. Psychopharmacol Comm 1:469–473, 1977.
52. Segal DS, Mandell AJ: Differential behavioral effects of hypothalamic polypeptides. In Prange AJ (ed): "The Thyroid Axis, Drugs, and Behavior." New York: Raven Press, 1974, pp 129–133.
53. Smith JR, LaHann RT, Chesnut RM, Carino MA, Horita A: Thyrotropin-releasing hormone: Stimulation of colonic activity following intracerebroventricular administration. Science 196:660–662, 1977.
54. Snyder SH: Brain peptides as neurotransmitters. Science 290:976–983, 1980.
55. Steiner H, Künzi H, Studer RO: Accumulation of tritiated thyrotropin releasing hormone in different organs, especially in the thyroid. Experientia 30:1096–1098, 1974.
56. Vogel JR, Beer B, Clody DE: A simple and reliable conflict procedure for testing anti-anxiety agents. Psychopharmacologia 21:1–7, 1971.
57. Vogel RA, Cooper BR, Barlow TS, Prange AJ Jr, Mueller RA, Breese GR: Effects of thyro-

tropin-releasing hormone on locomotor activity, operant performance and ingestive behavior. J Pharmacol Exp Ther 208:161–168, 1979.

58. Vogel RA, Frye GD, Wilson JH, Kuhn CM, Mailman RB, Mueller RA, Breese GR: Attenuation of the effect of punishment by thyrotropin releasing hormone: Comparisons with chlordiazepoxide. J Pharmacol Exp Ther 212:153–161, 1980.
59. Wei E, Loh H, Way El: Thyrotropin-releasing hormone and shaking behavior in the rat. Nature (London) 253:739–740, 1975b.
60. Wei E, Sigel SSR, Loh HH, Way EL: Central sites of naloxone precipitated shaking in the anesthetized morphine-dependent rat. J Pharmacol Exp Ther 195:480–487, 1975a.
61. Wilson IC, Prange AJ Jr, Lara PP, Alltop LB, Stikeleather RA, Lipton MA: TRH (Lopremone): Psychobiological responses of normal women. I. Subjective experiences. Arch Gen Psychiat 29:15–21, 1973.
62. Winokur A, Utiger RD: Regional distribution of thyrotropin releasing hormone in rat brain. Science 185:265–267, 1974.
63. Yarbrough GG: On the neuropharmacology of thyrotropin releasing hormone (TRH). Prog Neurobiol 12:291–312, 1979.
64. Yarbrough GG, Singh DK: Intravenous thyrotropin releasing hormone (TRH) enhances the excitatory actions of acetylcholine (ACh) on rat cortical neurons. Experientia 34:390, 1978.
65. Youngblood WW, Lipton MA, Kizer JS: TRH-like immunoreactivity in urine, serum, and extrahypothalamic brain: Non-identity with synthetic pyroglu-hist-pro-NH_2 (TRH). Brain Res 151:99–116, 1978.

The Role of Peptides and Amino Acids as Neurotransmitters, pages 117–125
© 1981 Alan R. Liss, Inc., 150 Fifth Avenue, New York, NY 10011

The Distribution of Some Identified Peptide-Containing Pathways in the Central Nervous System Including the Basal Ganglia

Earl A. Zimmerman

This review discusses some aspects of recent attempts to map specific peptidergic pathways in the mammalian nervous system by immunohistochemical methods. It is a monumental and exciting task. More than 20 different peptides are considered to be formed in specific pathways where they may be involved in neuronal communication [Hökfelt et al, 1980; Snyder, 1980] and many more are likely to be found. An explosion of new data on the peptides in the last decade is providing a tremendous expansion of the chemical anatomy of the brain begun a decade earlier in reference to the biogenic amines [Fuxe, 1965]. Experience has already taught that a complete and accurate distribution of any of the pathways has probably not yet been fully accomplished. False negative results are as large a problem in immunohistochemistry as are false positive ones. In the latter case, the determinants of a particular antibody can be discerned with adequate absorption controls, but there remains the important limitation of the immunological method which is the inability to know whether the peptide sequence visualized is the one in question or part of a different, possibly unknown, molecule (eg, Met-enkephalin vs its shared sequence in β-lipotropin). Put simply, one cannot do complete chemistry with immunological methods. On the other hand, present chemical methods cannot be used to localize peptides in situ. Both approaches are necessary to understand more fully a particular peptide in a particular brain region, although it is difficult to do in the same preparation.

Because of the limitations of immunohistochemistry discussed, most of the maps described must be considered preliminary and subject to further modification as new data are obtained. Advances in studies in biosynthesis in two different peptide-forming systems have resulted in additional probes for immunohistochemistry. The simultaneous localization of different parts and products of the precursors of ACTH (Pro-opiocortin) and vasopressin (Pro-pressophysin) has resulted in some assurance that certain hypothalamic neurons actually produce these peptides [Krieger et al, 1980; Zimmerman, 1981]. Discovery of prohormones for the other peptides will greatly assist their localization in the future. However, there is already enough data to indicate that each of the pathways on Table I is probably unique.

THE HYPOTHALAMIC BRAIN

At present, the two best-known peptide pathways in regards to biosynthesis happen to have all their cell bodies of origin in the hypothalamus and axonal connections to many other brain sites (Table I). Pro-opiocortin, or a precursor very similar to it, appears to be formed in neurons in the arcuate and around the ventromedial nucleus of the hypothalamus [Krieger et al, 1980]. That prohormone appears to be ultimately converted to α-melanocyte-stimulating hormone (α-MSH), and β-endorphin from the adrenocorticotropin, and β-lipotropin portions, respectively [Gramsch et al, 1980]. Specific antisera to all these and other parts of Pro-opiocortin localize the same hypothalamic neurons making a strong argument that these cells actually can produce α-MSH or β-endorphin (see [Zimmerman, 1981] for review). Recently, another more dorsal and lateral group of neurons were found which react with antiserum to α-MSH, but not with β-endorphin or other parts of Pro-opiocortin [Watson and Akil, 1980]. These neurons may contain some other unknown peptide which shares antigenic determinants with α-MSH.

PROJECTIONS OF THE α-MSH–β-ENDORPHIN SYSTEM

Fibers arising from the arcuate and nearby regions appear to innervate other neurons within and outside the hypothalamus. Nonreactive perikarya of the arcuate (Fig. 1) receive terminals where β-endorphin could act to stimulate prolactic secretion by inhibiting a dopaminergic arcuate neuron [Zimmerman and

TABLE I. Some Indentified Brain Peptide Pathways

Neuronal cell bodies relatively restricted to hypothalamus
ACTH–β-LPH forming α-MSH–β-endorphin
Pro-oxyphysin and Pro-pressophysin forming oxytocin, vasopressin, and respective neurophysins
Bradykinin
Luteinizing hormone releasing hormone
Peptides contained in perikarya in many regions
Substance P
Somatostatin
Neurotensin
Vasoactive intestinal polypeptide
Thyrotropin-releasing hormone
Methionine-enkephalin
Leucine-enkephalin

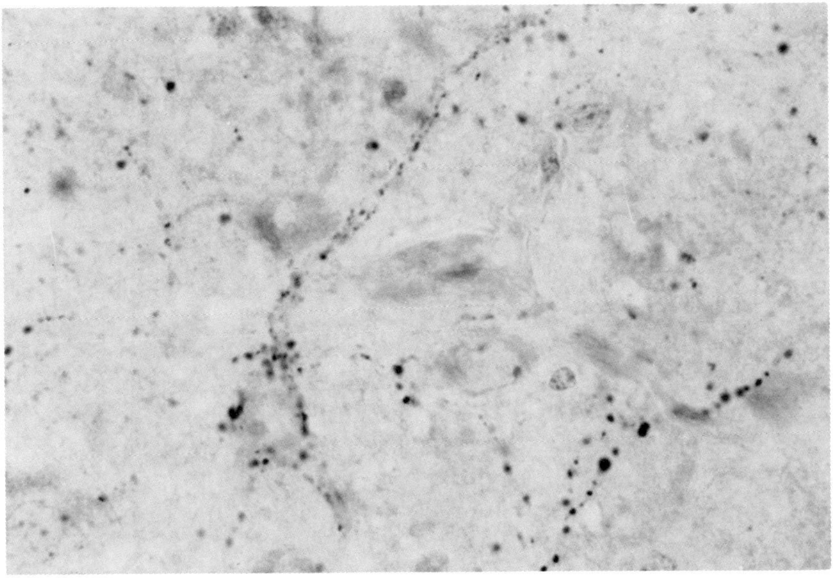

Fig. 1. Photomicrograph of arcuate (infundibular) nucleus of bovine hypothalamus reacted with antiserum to β-lipotropin and immunoperoxidase technique. Counterstain with cresyl violet. Some beaded reactive neurons innervated nonreactive neurons. Reprinted with permission from E.A. Zimmerman et al [Cell Tiss Res 186:393, 1978].

Nilaver, 1981]. A relatively dense innervation pattern is also seen in the periventricular and paraventricular nucleus, and there are moderate numbers of fibers elsewhere in the hypothalamus.

Extensive projections are found to other brain regions from the hypothalamus which are lost when the arcuate region is destroyed by the administration of monosodium glutamate to the neonatal rat [Krieger et al, 1980]. Notable areas which appear to receive fibers are the lateral septal nucleus, the bed nucleus of the stria terminalis, nucleus accumbens, amygdala, periventricular thalamus, periaqueductal grey, locus coeruleus, and nucleus of the solitary tract. Some fibers were also reported in substantia nigra (see [Krieger et al, 1980] for references). Few, if any, go to the spinal cord.

THE VASOPRESSIN–OXYTOCIN AND NEUROPHYSIN SYSTEM

In addition to its well-known neurosecretory projections to the posterior pituitary gland and hypophyseal portal system of the median eminence, recent evidence indicates that the magnocellular system of the hypothalamus has connections with other neurons outside of the hypothalamus [Swanson and Sawchenko, 1980]. Although both the supraoptic and paraventricular (PVN) nuclei produce oxytocin and vasopressin and their respective neurophysins and deliver them in granules to posterior pituitary nerve terminals, some of the cells of the PVN project to lateral septum, medial thalamus, amygdala, parabrachial nucleus, locus coeruleus, central grey, and solitary complex—areas overlapping the α-MSH–β-endorphine system, except for the spinal cord which is richly innervated by the PVN. Although both vasopressin and oxytocin appear to be transported to all these regions, the forebrain is thought to receive more vasopressin, while the hindbrain tends to contain more oxytocin fibers [Swanson and Sawchenko, in press; Buijs and Swaab, 1979; Nilaver et al, 1980]. That these peptides may serve as neurotransmitters was supported by their demonstration in granules in the synaptic terminal in the forebrain [Buijs and Swaab, 1979]. The functions of this system are not fully known, but it seems likely that the PVN may participate in memory functions [Wimersma et al, in press] by direct connections to the forebrain, in pain by fibers to midbrain central grey or substantia gelatinosa of spinal cord, and in autonomic regulation in solitarius, locus coeruleus, or intermediolateral grey of spinal cord [Swanson and Sawchenko, 1980]. Fibers to substantia nigra will be discussed later.

In reference to the other peptides listed under the "hypothalamic brain," luteinizing hormone-releasing hormone (LH-RH) is relatively restricted to the preoptic area and hypothalamus [Snyder, 1980], but perikarya extend into the olfactory regions as well [A-J Silverman and M. Fukuda, personal communication]. The initial report on the distribution of bradykinin by immunohistochemistry [Correa et al, 1979] suggests that all the cell bodies of origin may also lie in the hypothalamus (dorsomedial and periventricular nuclear regions). Fibers were found in the hypothalamus, perirhinal and cingulate cortex, lateral septum, ventral caudate-putamen, and central grey.

PEPTIDES WITH CELL BODIES IN MANY REGIONS

The seven peptides on this list (Table I) are fairly well characterized and have been recently reviewed in detail [Hökfelt et al, 1980; Snyder, 1980]. Some general comments can be made. They all tend to be concentrated in limbic brain, brainstem autonomic centers, and dorsal horn of the spinal cord. They are all very concentrated in the hypothalamus, particularly in nerve fibers to median eminence, where they serve as releasing or inhibiting factors at the hypothalamic or anterior pituitary level. The innervation of this region by vasoactive intestinal polypeptide (VIP) is relatively poor, however, suggesting that it may not act at the pituitary level [Sims et al, 1980]. The amygdala is also usually heavily innervated. The fiber patterns for VIP, enkephalin, and in neurotensin are nearly identical in the central nucleus and the bed nucleus of the stria terminalis [Sims et al, 1980]. These three peptides, as well as substance P, are found in locus coeruleus and the A_2 region of the medulla [Hökfelt et al, 1980]. Dendritic terminals containing substance P or enkephalin on catecholamine neurons at this site may play a role in the regulation of pain, motility, or blood pressure [Pickel et al, 1979]. In the spinal cord, most of the peptides are probably involved in one of several ways in pain mechanisms. Substance P, and somatostatin, and possibly VIP and cholecystokinin are located in primary sensory afferents [Hökfelt et al, 1980]. Neurotensin and enkephalin are found in interneurons in the dorsal horn [Hökfelt et al, 1980].

VIP, the octapeptide c-terminal form of cholecystokinin [Sims et al, 1980], and somatostatin [Krisch and Leonhardt, 1980] are unique in being found in neuronal cell bodies in the cerebral cortex. Another general finding is the absence of all the peptides studied thus far in the cerebellum.

PEPTIDES IN BASAL GANGLIA

Eight different peptide systems (Table II) have been reported in the extrapyramidal system. It is well known that the heavy innervation of the substantia nigra by fibers containing substance P originate in the caudate-putamen [Hong et al, 1977; Cuello, 1978; Kanazarva et al, 1980] which appear to be stimulatory [Kelley and Iverson, 1978]. A deficiency in the substance P [Gale et al, 1978], as well as GABA, is found in patients with Huntington's Disease. Enkephalin cell bodies in caudate-putamen, in contrast, innervate globus pallidus [Cuello, 1978], whereas opiates may increase the turnover of GABA [Moroni et al, 1978]. Both leu- and met-enkephalin are formed in the striatum [Snyder, 1980]. β-Endorphin fibers to nigra from hypothalamus, mentioned earlier, may affect GABA at this site, although such fibers are few.

Fibers reactive to antiserum to cholecystokinin were reported to be distributed heavily in rat caudate in one study [Loren et al, 1979], but not in two others [Innis et al, 1979; Hökfelt et al, 1980]. Some cells containing VIP were recently found in caudate-putamen in our own material (Fig. 2), and some were noted in this area to react with antiserum to somatostatin [Krisch and Leonhardt, 1980].

TABLE II. Neuronal Peptides in Basal Ganglia

Peptide	Cell bodies	Fibers
Substance P	Caudate-putamen Globus pallidus	Substantia nigra Globus pallidus
Enkephalins	Caudate putamen	Globus pallidus
Cholecystokinin	?	Caudate-putamen
VIP	Caudate-putamen	—
Somatostatin	Caudate-putamen	Caudate
Bradykinin	Hypothalamus (DM)	Caudate, Globus pallidus
Oxyotcin	Hypothalamus (PVN)	Substantia nigra
β-Endorphin–α-MSH	Hypothalamus (ARC)	Substantia nigra

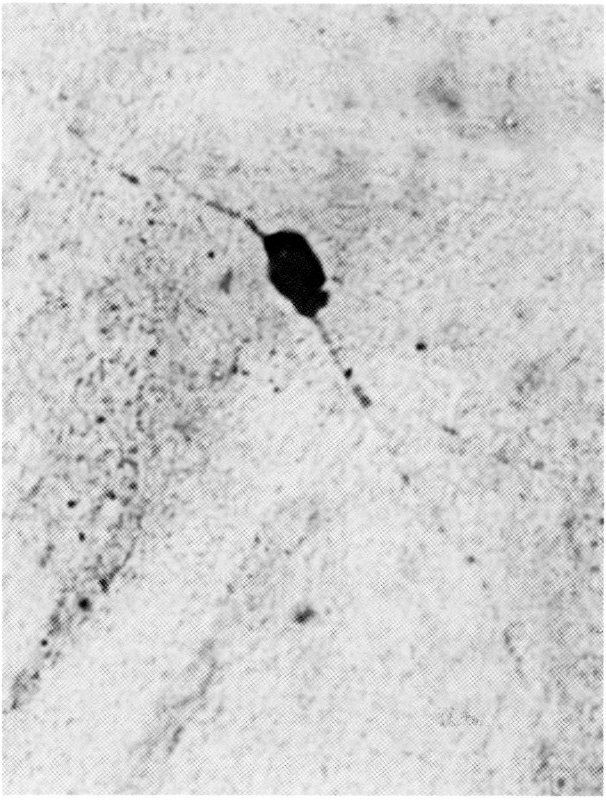

Fig. 2. VIP in a neuron in rat caudate-putamen by immunoperoxidase technique [D.L. Hoffman and E.A. Zimmerman, unpublished].

Further studies are needed to confirm and extend these findings. It will also be of interest to know in the future if the bradykinin fibers in caudate and pallidum originate in hypothalamus [Correa et al, 1979]. The hypothalamus may also play a role in the extrapyramidal system via PVN fibers to substantia nigra carrying oxytocin ([Nilaver et al, 1980] Fig. 3). If the oxytocin is converted to MSH-inhibiting hormone (MIF) it may act to potentiate dopaminergic neurons [Plotnikoff et al, 1971]. MIF was previously investigated as a possible treatment for Parkinson's Disease [Barbeau et al, 1976]. Interest in this possibility has been rekindled in our laboratories by a more potent analog (Pro-D-Leu-Gly-amide) parenterally active in rats [Snider et al, 1980].

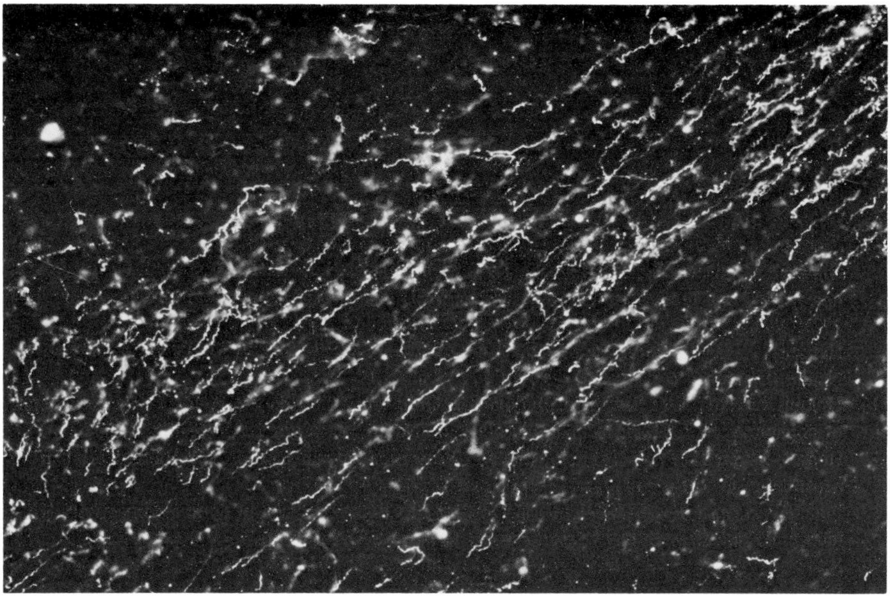

Fig. 3. Neurophysin-reactive fibers in rat substantia nigra. Coronal section, right side. Darkfield micrograph, immunoperoxidase technique. The main descending fiber pathway from the paraventricular nucleus of the hypothalamus in the midbrain passes through the pars compacta of the substantia nigra (central portion of photomicrograph). Some fibers extend into pars reticulata and probably innervate this regions (lower right hand corner) [G. Nilaver, A. Starr, and E.A. Zimmerman, unpublished].

Unfortunately, we know very little about the real role of the peptides discussed in reference to basal ganglia or other brain regions as well. The systems are only partially described and many more will be found. From what is known, however, it seems certain that peptide-specific diseases will be found and rational new therapies will emerge from basic research in the neuropeptides.

REFERENCES

1. Barbeau A, Roy M, Kastin AJ: Double blind evaluation of oral l-prolyl-leucyl-glycine amide in Parkinson's Disease. Canad Med Assoc J 114:120–122, 1976.
2. Bodnar RJ, Zimmerman EA, Nilaver G, Mansour A, Thomas LW, Kelly DD, Glusman M: Dissociation of cold-water swim and morphine analgesia in Brattleboro rats with diabetes insipidus. Life Sci 26:1581–1590, 1980.
3. Buijs RM, Swaab DF: Immuno-electron microscopical demonstration of synapses in the limbic system of the rat. Cell Tiss Res 204:355–365, 1979.

4. Correa FMA, Innis RB, Uhl G, Snyder S: Bradykinin immunoreactive systems localized histochemically in rat brain. Proc Nat Acad Sci 76:1489–1493, 1979.
5. Cuello AC: Enkephalin and substance P containing neurons in the trigeminal and extrapyramidal systems. In Costa E and Trabucchi M (eds): "Advances in Biochemical Pharmacology, Vol 18." New York: Raven Press, 1978, pp 111–123.
6. Fuxe K: Evidence of the existence of monoamine terminals in the central nervous system. Acta Physiol Scand (Suppl 247) 64:37–85, 1965.
7. Gale JS, Brd ED, Spokes EG, Iverson LL, Jessel T: Human brain substance P: Distribution in controls and Huntington's Chorea. J Neurochem 30:633–634, 1978.
8. Gramsch C, Kleber G, Hollt V, Pasi A, Mehraein P, Herz A: Pro-opiocortin fragments in human and rat brain: β-Endorphin and α-MSH are the predominant peptides. Brain Res 192:109–119, 1980.
9. Hökfelt T, Johansson O, Ljungdahl A, Lundberg JM, Schultzberg M: Peptidergic neurons. Nature (London) 284:515–521, 1980.
10. Hökfelt T, Rehfeld JF, Skirboll L, Ivemark B, Goldstein M, Markey K: Evidence for coexistence of dopamine and CCK in mesolimbic neurons. Nature (London) 285:476–478, 1980.
11. Hong JS, Yang H-YT, Racagni G, Costa E: Projections of substance P-containing neurons from neostriatum to substantia nigra. Brain Res 122:541–544, 1977.
12. Innis RB, Correa FMA, Schneider B, Snyder S: Cholecystokinin-like immunoreactivity: Histochemical localization in rat brain. Proc Nat Acad Sci 76:521–525, 1979.
13. Kanazawa I, Mogaki S, Muramoto O, Kuzuhara S: On the origin of substance P-containing fibers to the entopeduncular nucleus and the substantia nigra of the rat. Brain Res 184:481–485, 1980.
14. Kelley AE, Iverson SD: Behavioral response to bilateral injections of substance P into the substantia nigra of the rat. Brain Res 158:474–478, 1978.
15. Krieger DT, Liotta AS, Brownstein MJ, Zimmerman EA: ACTH, β-liptropin, and related peptides in brain, pituitary and blood. Recent Prog Horm Res 36:277–344, 1980.
16. Krisch B, Leonhardt H: An intermittent somatostatin immunoreactivity in the cortex and basal ganglia of the rat. Cell Tiss Res 205:327–331, 1980.
17. Loren I, Alumets J, Hakanson R, Sundler F: Distribution of gastrin and CCK-like peptides in rat brain. Histochemistry 59:249–257, 1979.
18. Moroni P, Cheney DL, Peralta E, Costa E: Opiate receptor agonists as modulators of λ-aminobutyric acid turnover in the nucleus caudatus, globus pallidus and substantia nigra of the rat. J Pharmacol Exp Therapeut 207:870–877, 1978.
19. Nilaver G, Zimmerman EA, Wilkins J, Michaels J, Hoffman DL, Silverman A-J: Magnocellular projections to the lower brainstem and spinal cord of the rat. Neuroendocrinology 30:150–158, 1980.
20. Pickel VM, Joh TH, Reis D, Leeman SE, Miller RJ: Electronmicroscopic localization of substance P and enkephalin in axon terminals related to dendrites of catecholamine neurons. Brain Res 160:387–400, 1979.
21. Plotnikoff NP, Kastin AJ, Anderson MS, Schally AV: DOPA potentiation by hypothalamic factor, MSH-release inhibiting hormone (MIF). Life Sci 10:1279–1283, 1971.
22. Sims KB, Hoffman DL, Said SI, Zimmerman EA: Vasoactive intestinal polypeptide (VIP) in mouse and rat brain: An immunocytochemical study. Brain Res 186:165–183, 1980.
23. Snider SR, Matthew E, Roberts P, Zimmerman EA, Cort JH, Hollander S: Potentiation of Levodopa-induced behavior by a synthetic MIF (pro-leu-gly-NH2) analog in two experimental models of Parkinsonism. Neurology 30:366 (abstract), 1980.
24. Snider S: Brain peptides as neurotransmitters. Science 209:976–983, 1980.

25. Swanson LW, Sawchenko PE: The paraventricular nucleus: A site for the integration of neuroendocrine and autonomic mechanisms. Neuroendocrinology 31:410–417, 1980.
26. Watson SJ, Akil H: α-MSH in rat brain: Occurrence within and outside of β-endorphin neurons. Brain Res 182:217–223, 1980.
27. Wimersma Greidanus Tj B Van: Neurohypophysial hormones. Their role in endocrine function and behavioral hemostasis. In Nemeroff CB and Dunn AJ (eds): "Behavioral Neuroendocrinology." Jamaica, NY: Spectrum Publishing, Inc., (in press).
28. Zimmerman EA: The organization of oxytocin and vasopressin pathways. In Martin JB, Reichlin S, and Bick K (eds): "Neurosecretion and Brain Peptides: Implication for Brain Function and Neurologic Disease." (Advances in Biochemical Psychopharmacology) New York: Raven Press, 1981, 63–75.
29. Zimmerman EA, Nilaver G: Neuroendocrinology. In Appel SH (ed): "Current Neurology." Boston: Houghton Mifflin, 1981, 356–376.

The Role of Peptides and Amino Acids as Neurotransmitters, pages 127–141
©1981 Alan R. Liss, Inc., 150 Fifth Avenue, New York, NY 10011

Acetylcholine as Modulator of Amino-Acid-Mediated Synaptic Transmission

K. Krnjević

Acetylcholine (ACh), though long considered likely to be an important central synaptic transmitter [Dale, 1938; Feldberg, 1945, etc], has proven to have actions on neurons that are more consistent with a modulatory rather than a transmitter role at most sites that have been investigated in the CNS. With the exception of Renshaw cells in the spinal cord—and possibly some thalamic neurons—which show rapid excitatory effects of ACh [Eccles et al, 1954; Curtis and Eccles, 1958; Phillis et al, 1967], elsewhere the excitatory action of ACh has consistently been relatively slow in onset and prolonged in duration with other characteristics of a modulatory function [Krnjević, 1975].

Although modulation is less readily defined than transmission, it is useful to consider separately chemical actions that appear to influence particularly the efficiency of other synapses. A transmitter is usually held to convey a relatively quick excitatory or inhibitory message across the synaptic gap, the time scale of operation being in milliseconds [Krnjević, 1974]. The most obvious example is cholinergic transmission (nicotinic) at the vertebrate neuromuscular junction.

Central synapses typically do not show simple, all-or-none transmission comparable to the cholinergic action in muscle, but it is reasonable to consider synaptic potentials lasting 10–100 msec as belonging to the same category of transmitter action. It is significant that one of the best-defined central excitatory mechanisms is that of nicotinic synapses on Renshaw cells. Other probable excitatory transmitters that operate very quickly include L-glutamate and L-aspartate. We are on even surer ground in classifying gamma-aminobutyric acid (GABA) and glycine as inhibitory transmitters; GABA in particular appears to be the mediator at most inhibitory synapses in the brain. Various features of the actions of the excitatory and inhibitory amino acids are discussed in other papers presented at this conference [Johnson and McLennan]. The evidence outlined below strongly indicates that a major function of ACh as a muscarinic agent in

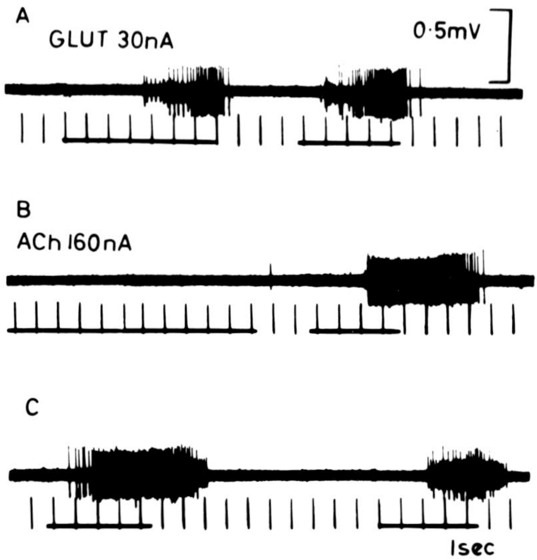

Fig. 1 Strong potentiation and prolongation of glutamate-evoked discharges by subthreshold doses of ACh. (A) Two units excited by two applications of glutamate (time of glutamate release is indicated by lines below time signals). (B) Last phase of 30-sec application of ACh. Subsequent tests of glutamate in (B) and (C) (which are continuous) show greatly enhanced response (Fig. 2 from [Krnjević et al, 1971b]).

the CNS is to facilitate excitatory synapses and to depress inhibitory synapses, in both cases the net result being enhanced postsynaptic responsiveness.

Postsynaptic Muscarinic Facilitation

A slow muscarinic excitatory effect of ACh was first observed in the mammalian cerebral cortex [Krnjević and Phillis, 1963a,b; Spehlmann, 1963]. An important characteristic of this action is that it facilitates, and especially prolongs neuronal responses evoked by dicarboxylic amino acids (Fig. 1) or by synaptic inputs that are probably not cholinergic. The postsynaptic site of facilitation was confirmed by later observations that cell firing evoked by intracellular current injections are also facilitated by ACh (Fig. 2) [Krnjević et al, 1971b; Woody et al, 1978; Dodd et al, 1981].

Comparable muscarinic excitatory and facilitatory actions have been seen in most regions of the brain and spinal cord (for a review, see [Krnjević, 1975]), and they are reflected in a wide distribution of muscarinic receptors, demonstrated in binding studies in the last few years [Kuhar and Yamamura, 1975; Birdsall et al, 1980 etc].

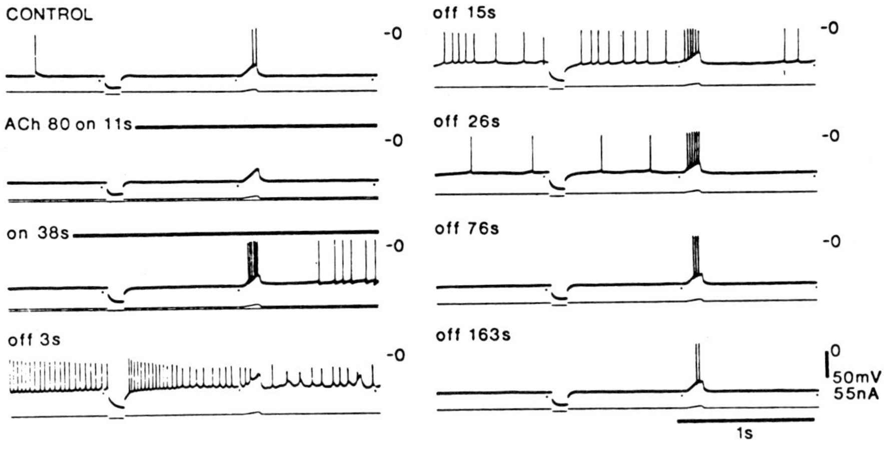

Fig. 2 Typical response of hippocampal neuron to iontophoretic application of ACh. Main traces show voltage response of cells to hyperpolarizing rectangular pulses and depolarizing ramps of current applied through intracellular microelectrode, just before and at various intervals during and after release of ACh (for 100 sec). After 38 sec of ACh release, there was a gradual depolarization (upward movement of trace), increase in number of spikes elicited by depolarizing ramps and continuous firing began. Note also increased response to hyperpolarizing pulses indicated a higher membrane resistance (Fig. 2 from [Dodd et al, 1981]).

Muscarinic Postsynaptic Facilitation Is Mediated by Reduction in K Conductance

Intracellular observations. In contrast to the nicotinic action at the muscle end plate, the central muscarinic depolarizing influence is associated with an increase in membrane resistance (Figs. 2 and 3). The original observations by Krnjević et al [1970, 1971b] on neocortical neurons have been confirmed by several groups of authors, studying cells in the neocortex [Woody et al, 1978], the hippocampus [Kelly et al, 1979; Dodd et al, 1981], the caudate nucleus [Bernardi et al, 1976], the spinal cord [Zieglgansberger and Reiter, 1974] and the interpeduncular nucleus [Sastry, 1980].

Depolarization accompanied by a higher resistance indicates a diminution in the membrane conductance for an ion having an equilibrium potential more negative than the resting potential. It could, therefore, be caused by a reduction in Cl^- or K^+ conductance. The depolarizing effect is observed even when IPSPs are reversed by intracellular Cl^- injection [Krnjević et al, 1971b]. Under these conditions, the Cl^- equilibrium potential is 30–50 mV more positive than the resting potential (V_m), and the reversal potential for the action of ACh (calculated from the intersection of the appropriate current–voltage lines) remains at a more

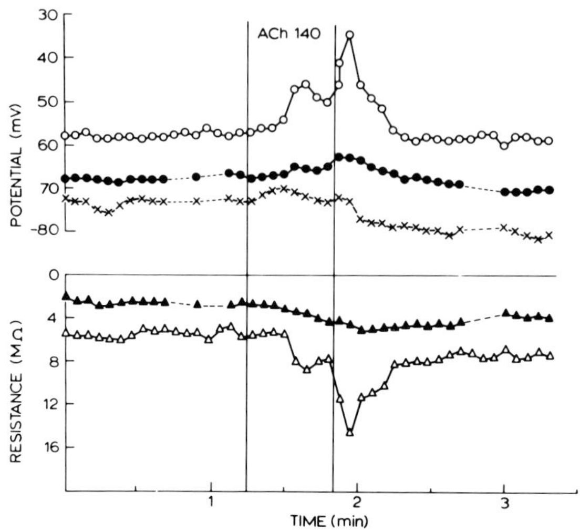

Fig. 3 Time course of depolarization and resistance increase produced by application of ACh to neocortical cell in cat. Membrane potential at rest is indicated by open circles, during inhibitory post-synaptic potentials (IPSPs) by full circles. Crosses are IPSPs reversal level. Input resistance is indicated, at rest by open triangles, and during IPSP by closed triangles (Fig. 10 from [Krnjević et al, 1971b]).

negative level than V_m and, indeed, is close to the expected value of E_K. It was, therefore, concluded that ACh reduces the K conductance (G_K), thus enhancing the net inward current generated by tonic or phasic excitatory influences mediated by other synaptic inputs.

Mechanism of reduction of G_K by ACh. In recent years, there has been increasing evidence for the existence of several kinds of membrane K^+ channels. In addition to the now "classical" voltage-dependent G_K involved in the repolarization of the action potential [Hodgkin, 1964], other kinds of voltage-dependent G_K have been reported [Connor and Stevens, 1971; Thompson, 1977; Brown and Adams, 1980], as well as a G_K that is triggered by intracellular Ca^{2+} [Krnjević et al, 1972, 1978; Meech, 1972, 1978]. Which of these various K^+ currents is (or are) switched off by ACh?

The best-known voltage-dependent G_K is very predictably blocked by tetraethylammonium (TEA) (and some related agents), which thus can greatly prolong the duration of action potentials, by preventing the efflux of K^+. As a close relative of TEA, and one that "looks" sufficiently like K^+ to be indistinguishable from the latter by some K^+-selective exchange resins [Neher and Lux, 1973], ACh might be expected also to occlude the Hodgkin–Huxley K^+ channels. And, indeed, injections of ACh into spinal motoneurons prolong the action potential

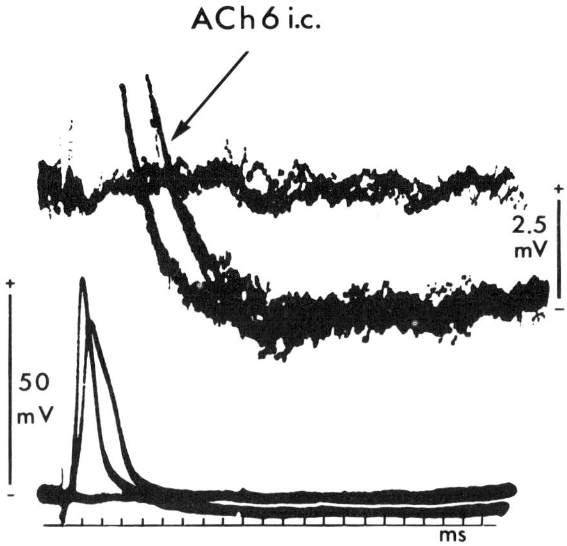

Fig. 4 Intracellular injection of ACh (6 nA for 1 min) selectively depresses fast outward current in cat spinal motoneuron. Superimposed traces of antidromic action potential and after hyperpolarization, before and after ACh injection, at two different gains below and above (from Fig. 10 in [Krnjević, 1977]).

about as effectively as do injections of TEA (Fig. 4) [Krnjević et al, 1976]. Though potentially of functional significance, this effect is probably not responsible for the main features of the muscarinic excitatory or facilitatory action, since intracellular ACh (or TEA) does not initiate either a slow depolarization or enhancement of firing.

An intriguing possibility is that variations in the cytoplasmic ACh level in cholinergic nerve terminals could provide an auto-regulation of ACh release. It is well known that the wider action potentials produced by blockers of fast K channels greatly enhance transmitter release (eg [Dunant et al, 1980]). Therefore, one might expect that a build-up of intraterminal ACh would generate wider action potentials which liberate greater amounts of ACh, until the cytoplasmic ACh falls to a lower level, and so on.

The possibility that ACh might act by diminishing the Ca^{2+}-activated G_K has been discussed earlier [Krnjević, 1977b]: ACh could have such an effect either by causing a reduction in cytoplasmic free Ca^{2+}—for example, by reducing Ca^{2+} influx, increasing Ca^{2+} sequestration or outward transport, or by reducing the Ca^{2+}-sensitivity of the K^+ channels. At present there is no evidence that such changes occur, and one can only consider this as an interesting possibility.

On the other hand, there is now excellent evidence that in sympathetic ganglion cells—where a comparable muscarinic action of ACh also generates a slow depolarization associated with a reduction in G_K [Weight et al, 1979]—muscarine selectively blocks a voltage-dependent K^+ current ("M-current") that is activated by relatively small depolarizations [Brown and Adams, 1980]. The existence of similar muscarine-sensitive voltage-dependent K^+ channels in central neurons would very satisfactorily account for their behavior in the presence of ACh. If such channels are indeed present, they would very effectively tend to limit excitation by depolarizing inputs. Therefore, inactivation of these channels by ACh would remove this breaking action and strongly facilitate excitation and especially repetitive firing.

It is significant that cortical cells and sympathetic ganglion cells both tend to be depolarized by Ba^{2+} [Krnjević et al, 1971a; Constanti et al, 1980], probably because in both cases Ba^{2+} blocks comparable muscarine-sensitive K^+ channels.

Conclusion. The muscarinic facilitation of central neurons appears to be a widespread mechanism of modulation that operates through a selective reduction in K^+ outward currents, and which is potentially available wherever cholinergic terminals are present to release ACh. One such putative cholinergic modulation is "cortical arousal," a phenomenon believed to be involved in consciousness, though still incompletely understood [Magoun, 1958; Spehlmann and Smathers, 1974; Krnjević, 1975; Steriade, 1978].

Muscarinic Disinhibition

Another kind of facilitating action of ACh also appears to operate in the hippocampus, and possibly other areas of the brain.

Extracellular observations. If ACh is released from a micropipet in (or very near) the pyramidal cell layer of areas CA1 or CA3, one sees a striking change in the response evoked by low-frequency fimbrial stimulation. In area CA1, where antidromic spikes are not generated [Andersen et al, 1973], the predominantly positive field potential, that reflects strong IPSPs generated in the pyramidal cells [Andersen et al, 1964], is usually diminished, and large, negative population spikes appear [Krnjević et al, 1980a]. As illustrated in Figure 5 this excitatory effect is mediated principally (though not exclusively) by muscarinic receptors—note even stronger effect of acetyl-β-methylcholine (methacholine), and no excitation by butyrylcholine a nicotinic agent—and it is comparable to the excitation produced by the GABA antagonist bicuculline.

Can this effect of ACh be explained solely by the postsynaptic cholinergic facilitation discussed above, which is known to occur also in hippocampal cells [Straughan, 1975; Spencer et al, 1978; Kelly et al, 1979; Dodd et al, 1981]? The answer is probably no, for the following reasons:

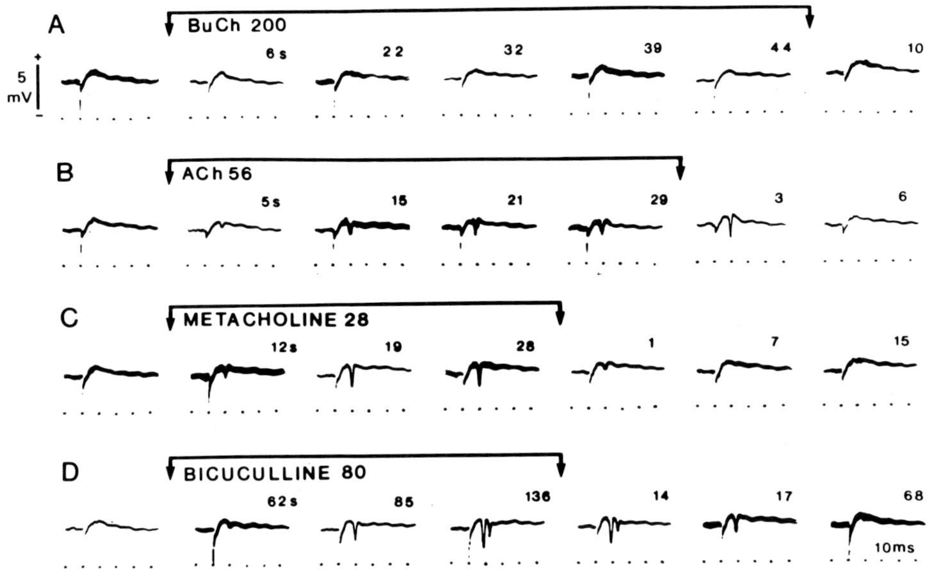

Fig. 5 Disinhibitory action of ACh in hippocampus. Extracellular field potentials were recorded with multibarrelled micropipette or from pyramidal layer in CA1 area of rat under urethane. Population spikes were evoked by iontophoretic release of ACh (B) and acetyl-β-methylcholine (C), but not by butyrylcholine, a nicotinic agonist (A), even though applied by much larger current. Bicuculline also evoked spikes after long latency (D). Times of recording after start and end of applications are indicated [K. Krnjević et al, unpublished observations].

(1) This action of ACh is very sharply localized to the pyramidal stratum.

(2) No comparable population spikes are evoked by L-glutamate, which is a very potent excitant of neurons at all levels in the hippocampus [Straughan, 1975; Spencer et al, 1978].

(3) It is remarkably similar to the effects produced by procedures or drugs that diminish the effectiveness of synaptic inhibition in the hippocampus:

- (a) higher frequency repetitive stimulation of the fimbria [Ben-Ari et al, 1979];
- (b) local applications of bicuculline (Fig. 5D);
- (c) local applications of MET-enkephalin, an endogenous opiate [Zieglgansberger et al, 1979; Dunwiddie et al, 1980; Krnjević et al, 1980b].

(4) Neither ACh nor repetitive stimulation can evoke population spikes if the intensity of fimbrial stimulation is relatively weak.

All these points are most consistent with a disinhibitory action of ACh: notably, item (1) because inhibitory terminals are especially concentrated on the pyramidal cell bodies [Andersen et al, 1964; Ribak et al, 1978], and item (4)

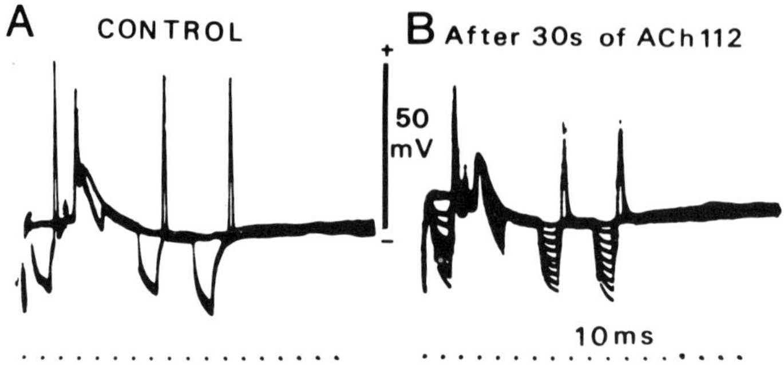

Fig. 6 Disinhibitory action of acetylcholine in hippocampus reduces efficiency of IPSPs. Initial intracellular records (A) show large positive IPSPs (due to Cl^- leakage from microelectrode) associated with marked fall in resistance, especially at peak of IPSP. During application of ACh (B), there was some depolarization (upward shift of base line), an increase in background resistance and a diminished fall during IPSP. Identical series of hyperpolarizing current pulses were injected into the cell at four different times to measure changes in resistance.

because disinhibition can become effective only if there is a sufficiently strong inhibitory input present in the first place.

Intracellular observations. Intracellular recordings from hippocampal cells support the above conclusion. Systematic observations of inhibitory postsynaptic potentials (IPSPs) evoked in these cells by fimbrial or entorhinal stimulation [Krnjević et al, 1980a] have shown a clear depressant effect of ACh, especially on the conductance increase that characterizes these IPSPs. This effect is illustrated in Figure 6. A characteristic feature of this action of ACh is its quick reversibility (within a few seconds), in contrast to the muscarinic facilitating effect which typically persists for 1 min (or even longer) after the end of the application of ACh (Figs. 1 and 2).

Mechanism of muscarinic disinhibition. The postulated depression of synaptic inhibition could operate in one of at least two ways:

(1) By analogy with the action of bicuculline (and other anti-GABA agents), ACh could interfere with the action of the inhibitory transmitter. Such a possible postsynaptic effect could be due to a selective block of Cl^- channels activated during inhibition [Kandel et al, 1961; Eccles et al, 1977] or by antagonism of the inhibitory (presumably GABA) receptors. Those are probably the least likely mechanisms, because as illustrated in Figure 7, ACh causes no detectible depression of the conductance increase elicited by direct application of GABA [Krnjević et al, 1980a; Ben-Ari et al, 1981].

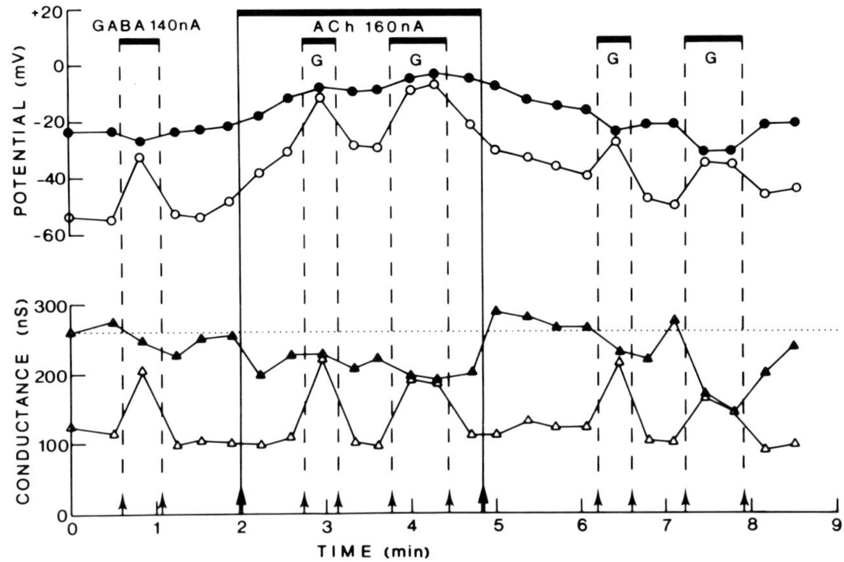

Fig. 7 Although ACh depresses hippocampal IPSPs, it does not reduce effectiveness of GABA. Data obtained from CA1 pyramidal cell in rat under urethane. IPSP was obtained throughout by stimulating fimbria at 0.6Hz, and GABA and ACh were applied at times indicated. Open circles, resting potential; black circles, IPSP potential; open triangles, resting input conductance; closed triangle, conductance at peak of IPSP.

(2) ACh could reduce inhibition by a presynaptic action, either by inhibiting the discharge of inhibitory interneurons or by diminishing the amount of transmitter released by active inhibitory terminals.

There are cholinergic parallels for both types of presynaptic actions. Thus, the cells of the reticular nucleus, which are believed to inhibit the principal relay cells of the ventro-basal thalamus [Dingledine and Kelly, 1977], are themselves inhibited by ACh [Ben-Ari et al, 1976]. Another possible example are the non-pyramidal cortical cells that fire in a characteristic bursting pattern at the onset of sleep and are likely to be inhibitory interneurons [Steriade, 1978]: such cells are sometimes strongly inhibited by ACh [Krnjević, 1974].

On the other hand, there is a good deal of evidence that ACh can depress transmitter release by an action on nerve terminals comparable to presynaptic inhibition of primary afferents. For example, an autoregulation of ACh release in the cortex has long been known [Mitchell, 1963], and it is now widely held to be caused by an action of ACh on cortical cholinergic terminals [Szerb et al, 1977]. The most relevant parallel observations are those of Hounsgaard [1978],

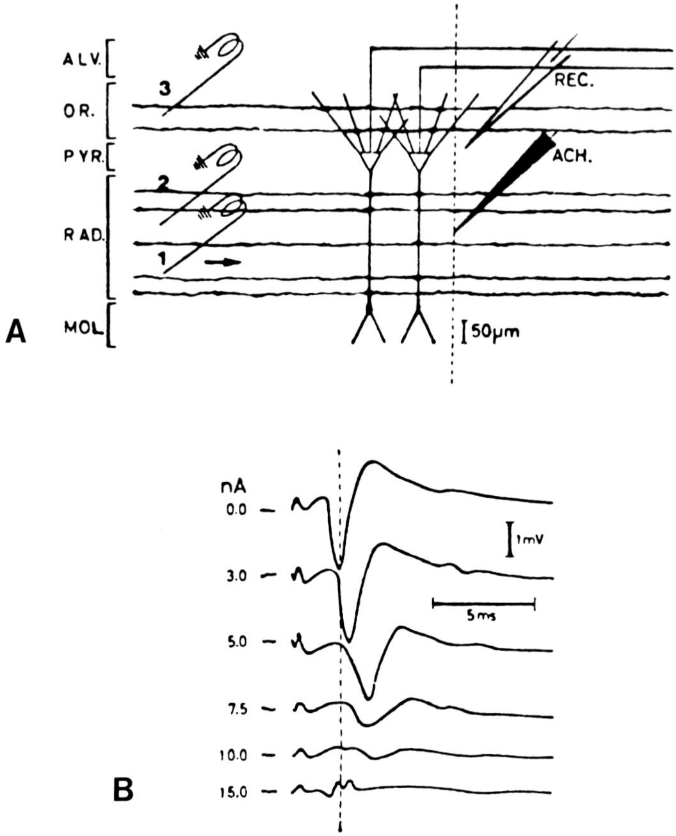

Fig. 8 ACh depresses release of transmitter at excitatory axo-dendritic synapses in hippocampal slice. (A) Diagram of area CA1 showing placement of electrodes and afferent fibers forming excitatory synapses with dendrites of pyramidal cells. (B) progressive depression of population spikes when increasing amounts of ACh were applied in synaptically activated region of dendrites (iontophoretic currents are indicated at left). Abbreviations: ACh, electrode for releasing ACh; Alv, alveus; MO1, stratum moleculare; Or, stratum oriens; Pyr, pyramidal cell layer; Rad, stratum radiatum; Rec, recording electrode. (part of Fig. 1 from [Hounsgaard, 1978]).

who obtained convincing evidence that ACh applied in dendritic regions of hippocampal slices strongly depresses excitatory synaptic transmission, by a presynaptic mechanism that appears to be associated with depolarization of the Schaeffer collateral nerve endings (Fig. 8). In view of Hounsgaard's [1978] findings, it seems likely that, by a similar kind of action, ACh can interfere with the release of the transmitter from the perisomatic inhibitory basket cell terminals.

Such a mechanism would help to explain why ACh evokes population spikes only when applied in the pyramidal layer.

Physiological significance of cholinergic disinhibition. The presence of cholinergic fibers in the hippocampus [Lewis et al, 1967; Dudar, 1975; Szerb et al, 1977] becomes especially significant in view of the juxta-pyramidal situation of their terminals [Storm-Mathisen 1977], where they would be best placed to influence the perisomatic inhibitory synapses.

The cells of origin of these cholinergic fibers have not been conclusively identified; but the greater weight of evidence favors their being situated mainly in the medial septal nuclei and deeper areas of the diagonal band [Lewis et al, 1967; Dudar, 1975; Storm-Mathisen, Mathisen, 1977; Lynch et al, 1978]. In preliminary experiments, stimulation of this region of the forebrain has, indeed, proven capable of evoking population spikes in the hippocampus, comparable to those produced at the same site by local iontophoresis of ACh [Ropert et al, 1980]. The observations are, therefore, consistent with the possibility that septo-hippocampal fibers can modulate hippocampal pyramidal excitability through a double facilitating mechanism: by a postsynaptic action tending to reduce K^+ conductance and a presynaptic disinhibition tending to diminish ongoing GABA release and, therefore, lower Cl^- conductance. As a result, there would be a marked depression of outward currents (of K^+ and Cl^-) that normally oppose depolarization. Hence, one can expect to see a particularly effective facilitation of pyramidal cell firing.

Conclusion. The double-edged facilitating (modulatory) action of ACh described here is likely to be a widespread mode of regulating neuronal activity in the CNS, judging by the wide distribution of ACh-evoked facilitating effects, though the relative contributions of postsynaptic and presynaptic (disinhibitory) mechanisms at various sites remain to be established.

There is no reason to believe that such facilitating modulatory mechanisms are exclusively cholinergic. It appears that some other types of slow and prolonged excitatory effects are also mediated by a postsynaptic inactivation of K^+ currents: eg, serotonergic facilitations in the myenteric plexus [Wood and Mayer, 1978; Johnson et al, 1980] and the facial motor nucleus [VanderMaelen and Aghajanian, 1980]; peptidergic actions like those of substance P, angiotensin and luteinizing hormone-releasing hormone [Krnjević, 1977; Adams and Brown, 1980; Jan et al, 1980]; and at least one endogenous opiate, MET-enkephalin, appears to have an ACh-like disinhibitory action in the hippocampus [Zieglgansberger et al, 1978; Dunwiddie et al, 1980; Krnjević et al, 1980b]. These findings support the view [Krnjević, 1974] that, like ACh, peptides more often operate as modulators of other—especially amino acid-mediated—synaptic transmissions, rather than as "classical," quick-acting neurotransmitters.

SUMMARY

(1) At many sites in the CNS, ACh as a muscarinic agent, facilitates neuronal firing, especially in response to other depolarizing inputs, whether mediated by synaptic action, the application of excitatory amino acids, or direct electrical stimulation.

(2) The facilitation acts in two different ways: by a direct, postsynaptic mechanism, and an indirect one, leading to disinhibition.

(3) The direct postsynaptic effect (seen for example in the cortex, hippocampus, spinal cord, etc) is produced by a reduction in membrane K conductance and, therefore, a depression of the outward K^+ current that would oppose depolarization and excitation. Recent observations on sympathetic ganglia suggest that ACh blocks selectively a voltage-dependent K^+ current ("M-current") that is activated by relatively small depolarization and would thus be particularly effective in preventing repetitive firing.

(4) There are indications that a variety of other endogenous substances can modulate cell firing through similar mechanisms (eg, monoamines and peptides).

(5) The second mechanism of cholinergic facilitation (seen especially well in the hippocampus) reduces the effectiveness of inhibitory synaptic activity, either by depressing the discharge of inhibitory interneurons or by reducing the liberation of GABA from inhibitory terminals (the second possibility seems more likely).

(6) At least one other endogenous substance, MET-enkephalin, also facilitates firing in the hippocampus by a comparable disinhibitory mechanism.

(7) These two kinds of modulatory actions of ACh both result in diminished responsiveness of some central neurons, owing to a depression of outward currents of K^+ or Cl^- that oppose excitation.

(8) Another possible intracellular modulatory action of ACh is suggested by the fact that intramotoneuronal injections of ACh increase the duration of action potentials (like TEA and other blockers of fast K^+ channels). Since prolonged action potentials induce a greater liberation of ACh from cholinergic terminals, variations in cytoplasmic levels of ACh in cholinergic terminals could provide a mechanism for autoregulation of ACh release.

REFERENCES

1. Adams PR, Brown DA: Luteinizing hormone-releasing factor and muscarinic agonists act on the same voltage-sensitive K^+ current in bullfrog sympathetic neurones. Br J Pharmacol 68:353–355, 1980.
2. Andersen P, Bland BH, Dudar JD: Organization of the hippocampal output. Exp Brain Res 17:152–168, 1973.
3. Andersen P, Eccles JC, Løyning Y: Location of postsynaptic inhibitory synapses on hippocampal pyramids. J Neurophysiol 27:592–607, 1964.
4. Ben-Ari Y, Dingledine R, Kanazawa I, Kelly JS: Inhibitory effects of acetylcholine on neurones in the feline nucleus reticularis thalami. J Physiol 261:647–671, 1976.

5. Ben-Ari Y, Krnjević K, Reinhardt W: Hippocampal seizures and failure of inhibition. Canad J Physiol Pharmacol 57:1462–1466, 1979.
6. Ben-Ari Y, Krnjević K, Reinhardt W, Ropert N: Intracellular observations on disinhibitory action of acetylcholine in hippocampus. Neuroscience (in press), 1981.
7. Bernardi G, Floris V, Marciani MG, Morocutti C, Stanzione P: The action of acetylcholine and L-glutamic acid on rat caudate neurons. Brain Res 114:134–138, 1976.
8. Birdsall NJM, Hulme EC, Burgen A: The character of the muscarinic receptors in different regions of the rat brain. Proc Roy Soc B 207:1–12, 1980.
9. Brown DA, Adams PR: Muscarinic suppression of novel voltage-sensitive K^+ current in a vertebrate neurone. Nature (London) 283:673–676, 1980.
10. Connor JA, Stevens CF: Voltage clamp studies of a transient outward membrane current in gastropod neural somata. J Physiol 213:21–30, 1971.
11. Constanti A, Adams PR, Brown DA: Why do barium ions imitate acetylcholine? Brain Res 206:244–250, 1981.
12. Curtis DR, Eccles RM: The excitation of Renshaw cells by pharmacological agents applied electrophoretically. J Physiol 141:435–445, 1958.
13. Dale HH: Acetylcholine as a chemical transmitter of the effects of nerve impulses. J Mt Sinai Hosp IV:401–429, 1938.
14. Dingledine R, Kelly JS: Brain stem stimulation and the acetylcholine-evoked inhibition of neurones in the feline nucleus reticularis thalami. J Physiol 271:135–154, 1977.
15. Dodd J, Dingledine R, Kelly JS: The excitatory action of acetylcholine on hippocampal neurones of the guinea pig and rat maintained *in vitro*. Brain Res 207:109–127, 1981.
16. Dudar JD: The effect of septal nuclei stimulation on the release of acetylcholine from the rabbit hippocampus. Brain Res 83:123–133, 1975.
17. Dunant Y, Eder L, Servetiadis-Hirt L: Acetylcholine release evoked by single or a few nerve impulses in the electric organ of Torpedo. J Physiol 298:185–203, 1980.
18. Dunwiddie T, Mueller A, Palmer M, Stewart J, Hoffer B: Electrophysiological interactions of enkephalins with neuronal circuitry in the rat hippocampus. I. Effects on pyramidal cell activity. Brain Res 184:311–330, 1980.
19. Eccles JC, Fatt P, Koketsu K: Cholinergic and inhibitory synapses in a pathway from motor-axon collaterals to motoneurones. J Physiol 126:524–562, 1954.
20. Eccles JC, Nicoll RA, Oshima T, Rubia FJ: The anionic permeability of the inhibitory post-synaptic membrane of hippocampal pyramidal cells. Proc R Soc Lond B 198:345–361, 1977.
21. Feldberg W: Present views on the mode of action of acetylcholine in the central nervous system. Physiol Rev 25:596–642, 1945.
22. Hodgkin AL: "The Conduction of the Nervous Impulse." Liverpool: Liverpool University Press, 1964, p 63.
23. Hounsgaard J: Presynaptic inhibitory action of acetylcholine in area CA1 of the hippocampus. Exp Neurol 62:787–797, 1978.
24. Jan YN, Jan LY, Kuffler SW: Further evidence for peptidergic transmission in sympathetic ganglion. Proc Nat Acad Sci 77:5008–5012, 1980.
25. Johnson SM, Katayama Y, North RA: Multiple actions of 5-Hydroxytryptamine on myenteric neurones of the guinea pig ileum. J Physiol 304:459–470, 1980.
26. Kandel ER, Spencer WA, Brinley FJ: Electrophysiology of hippocampal neurons. I. Sequential invasion and synaptic organization. J Neurophysiol 24:225–241, 1961.
27. Katayama Y, North RA, Williams JT: The action of substance P on neurons of the myenteric plexus of the guinea pig small intestine. Proc Roy Soc B 206:191–208, 1979.
28. Kelly JS, Dodd J, Dingledine R: Acetylcholine as an excitatory and inhibitory transmitter in the mammalian central nervous system. Prog Brain Res 49:252–266, 1979.
29. Krnjević K: Chemical nature of synaptic transmission in vertebrates. Physiol Rev 54:418–540, 1974.

30. Krnjević K: Acetylcholine receptors in vertebrate CNS. In Iversen LL, Snyder SH (eds): "Handbook of Psychopharmacology," Vol 6. New York: Plenum Press, 1975, pp 97–126.
31. Krnjević K: Effects of substance P on central neurons in cats. In von Euler US, Pernow B (eds): "Substance P." New York: Raven Press, 1977, pp 217–230.
32. Krnjević K: Control of neuronal excitability by intracellular divalent cations: A possible target for neurotransmitter actions. In Fields WS (ed): "Neurotransmitter Function: Basic and Clinical Aspects." Miami, Florida: Symposia Specialists Press, 1977, pp 11–26.
33. Krnjević K, Lisiewicz A: Injections of calcium ions into spinal motoneurones. J Physiol 225:363–390, 1972.
34. Krnjević K, Phillis JW: Acetylcholine-sensitive cells in the cerebral cortex. J Physiol 166:296–327, 1963a.
35. Krnjević K, Phillis JW: Pharmacological properties of acetylcholine-sensitive cells in the cerebral cortex. J Physiol 166:328–350, 1963b.
36. Krnjević K, Puil, E, Werman, R: EGTA and motoneuronal after-potentials. J Physiol 275:199–223, 1978.
37. Krnjević K, Pumain R, Renaud L: Excitation of cortical cells by barium. J Physiol 211:43–44P, 1970.
38. Krnjević K, Pumain R, Renaud L: Effects of Ba^{2+} and tetraethylammonium on cortical neurones. J Physiol 215:223–245, 1971a.
39. Krnjević K, Pumain R, Renaud L: The mechanism of excitation by acetylcholine in the cerebral cortex. J Physiol 215:247–268, 1971b.
40. Krnjević K, Reiffenstein RJ, Ropert N: Disinhibitory action of acetylcholine in the hippocampus. J Physiol 308:73–74P, 1980a.
41. Krnjević K, Reiffenstein RJ, Ropert N: Hippocampal disinhibition and endogenous opioids. The Physiologist 23:69, 1980b.
42. Kuhar MJ, Yamamura HI: Light autoradiographic localization of cholinergic muscarinic receptors in rat brain by specific binding of a potent antagonist. Nature (London) 253:560–561, 1975.
43. Lewis PR, Shute CCD, Silver A: Confirmation from choline acetylase analyses of a massive cholinergic innervation to the rat hippocampus. J Physiol 191:215–224, 1967.
44. Lynch G, Rose G, Gall C: Anatomical and functional aspects of the septo-hippocampal projections. In "Functions of the Septo-Hippocampal System," Ciba Foundation Symposium 58. Amsterdam: Elsevier, 1978, p 5–20.
45. Magoun HW: "The Waking Brain." Springfield: Thomas, 1958, p 138.
46. Meech RW: Intracellular calcium injection causes increased potassium conductance in *Aplysia* cells. Comp Biochem Physiol 42A:493–499, 1972.
47. Meech RW: Calcium-dependent potassium activation in nervous tissues. Ann Rev Biophys Bioeng 7:1–18, 1978.
48. Mitchell JF: The spontaneous and evoked release of acetylcholine from the cerebral cortex. J Physiol 165:98–116, 1963.
49. Neher E, Lux HD: Rapid changes of potassium concentration at the outer surface of exposed single neurons during membrane current flow. J Gen Physiol 61:385–399, 1973.
50. Phillis JW, Tebecis AK, York DH: A study of cholinoceptive cells in the lateral geniculate nucleus. J Physiol 192:695–713, 1967.
51. Ribak CE, Vaughn JE, Saito K: Immunocytochemical localization of glutamic acid decarboxylase in neuronal somata following colchicine inhibition of axonal transport. Brain Res 140:315–332, 1978.
52. Sastry BR: Excitatory post-synaptic potentials in the mammalian central nervous system associated with an increase in the membrane resistance. Life Sciences 27:1403–1407, 1980.
53. Spehlmann R: Acetylcholine and prostigmine electrophoresis at visual cortical neurons. J Neurophysiol 26:127–139, 1963.

54. Spehlmann R, Smathers Jr CC: The effects of acetylcholine and of synaptic stimulation on the sensorimotor cortex of cats. II. Comparison of the neuronal responses to reticular and other stimuli. Brain Res 74:243–253, 1974.
55. Spencer H, Gribkoff VK, Lynch GS: Distribution of acetylcholine, glutamate and aspartate sensitivity over the dendritic fields of hippocampal CAl neurones In Ryall RW, Kelly JS (eds): "Iontophoresis and Transmitter Mechanisms in the Mammalian Central Nervous System." Amsterdam: Elsevier/North-Holland Biomedical Press, 1978, pp 194–196.
56. Steriade M: Cortical long-axoned cells and putative interneurons during the sleep-waking cycle. Behav Brain Sci 3:465–514, 1978.
57. Storm-Mathisen J: Localization of transmitter candidates in the brain: The hippocampal formation as a model. Progr Neurobiol 8:119–181, 1977.
58. Straughan DW: Neurotransmitters and the hippocampus. In Isaacson RL, Pribram KH (eds): "The Hippocampus," Vol. 1: Structure and Development. New York: Plenum Press, 1975, pp 239–268.
59. Szerb JC, Hadhazy P, Dudar JD: Release of [^{3}H] acetylcholine from rat hippocampal slices: Effect of septal lesion and of graded concentrations of muscarinic agonists and antagonists. Brain Res 128:285–291, 1977.
60. Thompson SH: Three pharmacologically distinct potassium channels in molluscan neurones. J Physiol 265:465–488, 1977.
61. VanderMaelen CP, Aghajanian GK: Intracellular studies showing modulation of facial motoneurons excitability by serotonin. Nature (London) 287:346–347, 1980.
62. Weight FF, Schulman JA, Smith PA, Busis NA: Long-lasting synaptic potentials and the modulation of synaptic transmission. Federation Proc 38:2084–2094, 1979.
63. Wood JD, Mayer CJ: Slow synaptic excitation mediated by serotonin in Auerbach's plexus. Nature (London) 276:836–837, 1978.
64. Woody CD, Swartz BE, Gruen E: Effects of acetylcholine and cyclic GMP on input resistance of cortical neurons in awake cats. Brain Res 158:373–395, 1978.
65. Zieglgansberger W, French ED, Siggins GR, Bloom FE: Opioid peptides may excite hippocampal pyramidal neurons by inhibiting adjacent inhibitory interneurons. Science 205:415–417, 1979.
66. Zieglgansberger W, Reiter C: A cholinergic mechanism in the spinal cord of cats. Neuropharmacology 13:519–527, 1974.

The Role of Peptides and Amino Acids as Neurotransmitters, pages 143–149
©1981 Alan R. Liss, Inc., 150 Fifth Avenue, New York, NY 10011

Failure of Apomorphine to Induce Dopaminergic Hypersensitivity: A Replication

Abraham Flemenbaum

Cannon and Rosenblueth [1949] demonstrated the phenomenon of "denervation supersensitivity" consisting of an increased sensitivity of neuronal receptor to a specific transmitter when the afferents are interrupted. When the dopamine receptor is blocked for long periods of time, it becomes chemically denervated. This is followed by the development of supersensitivity in a manner similar to the one observed in Cannon's preparation. This supersensitivity provides a means by which the receptor can overcome the blocking agent and thus restore function to normal [Friedhoff, 1977]. Carlsson and Linquist [1963] showed that antipsychotic drugs block central dopamine receptors. Friedhoff and Van Winkle [1964] pointed out that all antipsychotic drugs were capable of producing a Parkinson-like syndrome, and as Parkinson is related to dopamine deficiency, they suggested that antipsychotic drugs may act by interferring with dopaminergic transmission. Carlsson on the one hand [1970] and Klawans et al [1970] on the other have proposed an analogous mechanism as an explanation for the appearance of Tardive Dyskinesia. In their theoretical model, the dyskinesia is produced by denervation hypersensitivity.

In the same vein, if blocking the receptor from the agonist produces supersensitization, then perhaps an increase of the agonist available to the receptor could produce subsensitization. Subsensitization in the form of decreased response at the receptor level or a reduction in the number of binding sites has been demonstrated to several transmitters including acetylcholine [Katz and Thesleff, 1957], insulin and glucagon [Goldstein et al, 1975], and thyroid-releasing hormone (TRH) [Hinkle et al, 1975].

As elegantly summarized by Friedhoff [1977], if the agonist supplies are excessive, the receptor is overstimulated. The receptor then compensates by decreasing its sensitivity, so when the agonist supply is returned to normal, the receptor is subsensitive. The reverse is also true, if the agonist supply is decreased, ie, from blockade of the receptor, there is a compensatory increase in receptor responsivity. However, the changes in receptor sensitivity set point appear to require intensive and prolonged treatment. This paradigm may be a general one referable to many types of receptors. The regulation of sensitivity may occur in a receptor cell or may result from an interaction and modulation of that cell by other cells including presynaptic neurons or synapsing inhibitory and/or excitatory neurons. Furthermore, modulation of the receptor sensitivity by agents other than the specific agonist may also occur.

Friedhoff [1977] then proposed a general hypothesis for modification of receptor sensitivity: (a) that receptor cells can change the receptor "set point," either increasing or decreasing the sensitivity; (b) these changes occur in the direction of restoration of homeostasis, so that a receptor blocker will provoke the cell to increase the sensitivity, whereas an oversupply of agonist will stimulate a reduction of the sensitivity; (c) that these processes proceed with a slow time course; and (d) that persistent changes in sensitivity reflect an enduring change in regulatory processes in the cell.

These observations provide an easy if simplistic explanation for some effects of the development of tolerance (progressive reduction of effects observed to the repetitive administration of the same dose of a drug) as well as for tardive and/or other dyskinesias. However, it does not explain that Parkinsonism can coexist in patients with Tardive Dyskinesia [Gerlach, 1977] or that a number of researchers have observed that repetitive administration of the same doses of amphetamine or cocaine to animals and humans produces increasing effects or "reverse tolerance" on some aspects of behavior [Ellinwood, 1967; Mago, 1969; Segal and Mandell, 1974; Ellinwood and Kilbey, 1975; Klawans and Margolin, 1975; Post and Kopanda, 1976; Post et al, 1976]. These changes in behavior have been observed specially in stereotyped behavior (SB), an animal model of psychosis closely related to dopamine (DA) receptor stimulation.

In Tardive Dyskinesia, a receptor chronically deprived of the neurotransmitter becomes hypersensitive to its presence. In the reverse-tolerance model, a receptor exposed to chronic stimulation, also increases its sensitivity. Klawans and Margolin [1975] called this phenomenon "agonist-induced hypersensitivity," suggesting what I call a "mirror image" of the pathophysiology of both denervation hypersensitivity and the agonist-induced hypersensitivity, ie, an increased sensitivity of the receptor exposed to chronic stimulation by the agonist, the same as the one exposed to chronic blocking.

A representative review of the literature showed that DA receptor hypersensitivity has been demonstrated in animals by sensitizing the receptors at pretest with indirect DA agonist, such as cocaine and amphetamines. None of the studies used a direct DA agonist such as apomorphine as the sensitizing drug. Apomorphine has been utilized, only at the retest, for the evaluation of the hypersensitivity produced after a sensitization period by other drugs [Flemenbaum, 1978]. Thus, it was necessary to demonstrate in the laboratory the presence or absence of the phenomenon of hypersensitivity in animals pretreated (sensitized) with apomorphine.

Utilizing our "drug abuse-kindling" model, we have tested Sprague–Dawley rats of both sexes for DA receptor hypersensitivity, with amphetamine-like drugs, but also with apomorphine both at pretest, and at post-test drug levels [Flemenbaum, 1977, 1978, 1979a,b,c].

To our surprise, we obtained results contrary to those observed with amphetamine and cocaine, ie, a lack of induction of DA receptor hypersensitivity in animals pretreated with apomorphine in contrair, there was a tendency to a decrease of SB at post-test. Although this decrease did not reach statistical significance, it suggested a subsensitivity of the receptors previously exposed to chronic intermittent doses of the direct DA agonist, apomorphine. Thus it is considered necessary to reevaluate the phenomenon of DA receptor sensitivity when apomorphine is both the sensitizing and the (post) test drug.

MATERIALS AND METHODS

For this duplication, we utilized a methodology similar but slightly modified to that one recently described by Klawans et al [1977], but we used Sprague–Dawley rats instead of guinea pigs, fifteen experimental males, five male controls, and five female controls of the same initial weight and age. Food and water were provided ad libitum.

Every day at the same time, 1300–1600 hr, the animals were placed in individual plastic cages and underwent the following schedule (see Flow Chart, Fig. 1): 1 week each of acclimation, simple handling, and the experimental procedure consisted of 1 week of subthreshold doses of apomorphine (0.25 mg/kg) given on Monday, Wednesday, and Friday. This was followed by 3 weeks of daily apomorphine 2.5 mg/kg for the experimental animals and the female controls, or saline for the male controls. Then after 1 week of drug lag, the animals were retested on the subthreshold apomorphine schedule previously described. Finally the animals were all exposed to 3 weeks of daily apomorphine 2.5 mg/kg. All the drugs were given ip at 1 ml/kg of body weight.

Stereotyped behavior was measured by an independent blind rater at 15-min intervals utilizing two standardized scales [Ellinwood and Kilbey, 1975; Klawans and Margolin, 1975], the first including both normal and SB, and the second measuring SB exclusively. A total daily score was obtained by adding all measures on both scales for the 2-hr experimental period.

The animals were considered to have shown SB if they were rated at least

Group	Sex	N	Week 1	2	3	4	5 - 6 - 7	8	9	10 - 11 - 12
1	M	15	ACCLIMATION	HANDLING	SALINE	SUBTHRESHOLD APOMORPHINE	Apomorphine 2.5 mg/kg	LAG	SUBTHRESHOLD APOMORPHINE	Apomorphine 2.5 mg/kg
2	F	5					Apomorphine 2.5 mg/kg			Apomorphine 2.5 mg/kg
3	M	5					Saline			Apomorphine 2.5 mg/kg

Fig. 1. Flow chart.

a 7 on Ellinwood–Kilbey's scale or a 2 on Klawans–Margolin scale for two or more consecutive rating periods (.5 hr).

RESULTS

Table I shows a significant (χ^2, 14.6 df 1, $P < 0.01$) decrease in the number of animals with a 7/2 rating or better, at post-test with subthreshold doses of apomorphine. This was true for the experimental males and both controls.

As there was no difference between the experimental and control animals, in spite of the significant decrease of animals showing SB at post-test with subthreshold apomorphine, we cannot conclude that apomorphine produces a desensitization of the receptors, but the results suggest again a failure of apomorphine to produce the DA receptor hypersensitivity as measures by SB.

DISCUSSION AND CONCLUSIONS

We are left with two contradictory observations. On the one hand (and in opposition to Friedhoff's Receptor Sensitivity Modification Model), indirect acting DA agonists produce a syndrome of increased SB at post-test that has been loosely termed agonist-induced receptor hypersensitivity. On the other hand, a

TABLE I. SB at Pre- and Post-test With Subthreshold Apomorphine*

	Pretest	Post-test
Experimental (n = 15)	12	1
Females (n = 5)	4	0
Male saline controls (n = 5)	5	1

*Number of animals with a 7/2 stereotyped behavior rating for ≥ 30 min pre- and post-test with subthreshold doses of apomorphine.

direct DA agonist does not. To reconcile these differences, speculation on the basis of representative literature in related topics becomes necessary.

A variety of stimulating drugs can induce a syndrome of hyperactivity with sniffing, gnawing, repetitive licking, biting, and abnormal behavior known as stereotyped behavior, which reportedly has a direct relationship with dopamine receptors (see summary in [Flemenbaum, 1978]). DA has also been implicated in the rotation behavior in rats after unilateral destruction of the nigrostriatal DA system [Ungerstedt, 1975]. However, indirect DA receptor agonist have effects other than at the dopaminergic system, and some so-called dopaminergic behaviors are influenced by neurotransmitters other than dopamine. Andén and Strömbom [1974] established that the rotational behavior mentioned above is influenced by serotonin and acetylcholine; Wolforth et al [1974] among others found supporting evidence for acetylcholine involvement in the physiology of dopamine-receptor response. This involvement can occur via feedback on the nigral neurons perhaps by gabaminergic striatal neurons [Ladinski et al, 1975; McGeer et al, 1974; Gerlach 1977]. Furthermore, DA hypersensitivity is not always accompanied by an increase of dopamine-stimulated adenylate cyclase activity [von Voightlander et al, 1973, 1975]. This suggests that supersensitivity may be mediated by mechanisms other than adenylate cyclase.

Costall and Naylor [1973, 1974] showed that some DA fibers relate to activity and some relate to SB. Lesions of ascending dopamine fibers of the lateral hypothalamus and/or pallidectomy resulted in hypokinesis. Lesions of ascending dopamine fibers to the mesolymbic system, tubercullum olfactorium, nucleus accumbens, and nucleus amigdaloideus resulted in decreased intensity of apomorphine SB. Thus, it is possible that receptor hypersensitivity at different levels may result in behaviors apparently opposite or contradictory.

Dextro- and levo- amphetamine and cocaine have effects on systems other than dopaminergic, ie, serotoninergic. Further evidence for a difference between the effects of amphetamine and cocaine and those of apomorphine, comes from a report by Angrist et al [1975] that most dopaminergic agonists such as amphetamine and others including piribedil, but excluding apomorphine, induce schizophreniform psychosis in humans, a fact also reported by Meltzer and Stahl [1976].

Our results are in contradiction to findings by Tarsy and Baldessarini [1974] that showed that apomorphine produced receptor supersensitivity, but in their case the supersensitivity was induced only after extended pretreatment with AMPT, reserpine, chlorpromazine, or haloperidol—all drugs that interfere with dopaminergic activation.

In conclusion: Chronic stimulation of DA receptors with indirect agonist of the amphetamine type has been consistently reported to produce DA receptor hypersensitivity. We were unable to demonstrate this receptor supersensitivity utilizing the direct DA receptor agonist apomorphine.

REFERENCES

1. Anden NE, Strömbom U: Adrenergic receptor blocking agents: Effects on central noradrenaline and dopamine receptors and on motor activity. Psychopharmacologia (Berlin) 38:91–103, 1974.
2. Angrist B, Thompson H, Shopsin B, Gerchon S: Clinical studies with dopamine-receptor stimulants. Psychopharmacologia 44:273–280, 1975.
3. Cannon WB, Rosenblueth A: "Supersensitivity of Denervated Structures." New York: McMillian, 1949.
4. Carlsson A: Biochemical implications of dopa induced action on the central nervous system with particular reference to abnormal movements. In Barbeau A, McDowell FH (eds): "L-dopa and Parkinsonism." Philadelphia: Davis, 1970, p 205.
5. Carlsson A, Lindqvist M: Effect of chlorpromazine or haloperidol on formation of 3-methoxytyramine and normetanephrine in mouse brain. Acta Pharmacol Toxicol 20:140–144, 1963.
6. Costall B, Naylor RJ: The role of telencephalic dopaminergic systems in the mediation of apomorphine-stereotyped behaviour. Eur J Pharmacol 24:8–24, 1973.
7. Costall B, Naylor RJ: The involvement of dopaminergic systems with the stereotyped behaviour patterns induced by methylphenidate. J Pharm Pharmacol 26:30–33, 1974.
8. Ellinwood EH Jr: Amphetamine psychosis: I. Description of the individuals and process. J Nerv Ment Dis 144:273–283, 1967.
9. Ellinwood EH Jr, Kilbey MM: Amphetamine stereotype: The influence of environmental factors and prepotent behavioral patterns on its topography and development. Biol Psychiatr 10:3–16, 1975.
10. Flemenbaum A: Dopamine receptors hypersensitivity: Further confirmation following drug abuse model. In Messiha F, Kenny A (eds): "Parkinson's Disease: Neurophysiological, Clinical and Related Aspects." New York: Plenum, 1977, pp 207–217.
11. Flemenbaum A: Postsynaptic supersensitivity and kindling: Further evidence of similarities. Am J Alcohol Abuse 5 (2):247–254, 1978.
12. Flemenbaum A: Rat dopaminergic hypersensitivity: II. Effects of sex. Neuropsychobiology 5 (4):222–231, 1979a.
13. Flemenbaum A: Rat dopamine hypersensitivity: I. Effect of age. Neuropsychobiology 5 (4):213–221, 1979b.
14. Flemenbaum A: Failure of apomorphine to induce dopamine receptor hypersensitivity Psychopharmacology (Berlin) 62:175–179, 1979c.
15. Friedhoff AJ: Receptor Sensitivity Modification (RSM) a new paradigm for the potential treatment of some hormonal and transmitter disturbances. Comp Psychiat 18:309–317, 1977.

16. Friedhoff AJ, Van Winkle E: Biological 0-methylation and schizophrenia. Psychiat Res Rep Am Psychiatr Assn 19:149–153, 1974.
17. Gerlach J: The relationship between parkinsonism and tardive dyskinesia. Am J Psychiatr 134:781–784, 1977.
18. Goldstein S, Blecher B, Binder R, Perrino P, Recant L: Hormone receptors 5. Binding of glucagon and insulin to human circulating mononuclear cells in diabetes mellitus. Endocrinol Res Commun 2:367–376, 1975.
19. Hinkle PM, Tashjian HA Jr: Thyrotropin-releasing hormone regulates the number of its own receptors in the GH_3 strain of pituitary cells in culture. Biochemistry 14(17):3845–3851, 1975.
20. Katz B, Thesleff S: A study of "desensitization" produced by acetylcholine at the motor end plate. J Physiol 138:63–80, 1957.
21. Klawans HL Jr, Margolin DI: Amphetamine-induced dopaminergic hypersensitivity in guinea pigs. Arch Gen Psychiatr 32:725–732, 1975.
22. Klawans H Jr, Ilahi MM, Shenker D: Theoretical implications of the use L-dopa in Parkinsonism. Acta Neurol Scand 46:409–441, 1970.
23. Klawans HL Jr, Weiner WJ, Nausieda PA: The effect of lithium in an animal model of tardive dyskinesia. Progress Neuro-Psychopharmacol 1:53–60, 1977.
24. Ladinsky H, Consolo S, Bianchi S, Samanin R, Ghezzi D: Cholinergic-dopaminergic interaction in the striatum the effect of 6-hydroxydopamine or pimozide treatment on the increased striatal acetylcholine levels induced by apomorphine, piribedil and d-amphetamine. Brain Res 84:221–226, 1975.
25. Magos L: Persistence of the effect of amphetamine on stereotyped activity in rats. Eur J Pharmacol 6:200–201, 1969.
26. McGeer PL, Grewaal DS, McGeer EG: Influence of noncholinergic drugs on rat striatal acetylcholine levels. Brain Res 80:211–217, 1974.
27. Meltzer HY, Stahl SM: The dopamine hypothesis of schizophrenia: A review. Schizo Bull 2:1:19–76, 1976.
28. Post RM, Kopanda RT: Cocaine, kindling and psychosis. Am J Psychiat 133:627–634, 1976.
29. Post RM, Kopanda RT, Black KE: Progressive effects of cocaine on behavior and central amine metabolism in rhesus monkeys: Relationship in kindling and psychosis. Biol Psychiatr 11; no. 4:403–419, 1976.
30. Segal DS, Mandell AJ: Long-term administration of d-amphetamine: Progressive augmentation of motor activity and stereotype. Pharmacol Biochem Behav 2:249–255, 1974.
31. Tarsy D, Baldessarini RJ: Behavioral supersensitivity to apomorphine following chronic treatment with drugs which interfere with the synaptic function of catecholamines. Neuropharmacology 13:927–940, 1974.
32. Ungerstedt U, Ljungbert T, Hoffer B, Siggins G: Dopaminergic supersensitivity in the striatum. Advan Neurol 9:57–65, 1975.
33. Von Voigtlander PF, Boukma SJ, Johnson GA: Dopaminergic denervation supersensitivity and dopamine stimulated adenyl cyclase activity. Neuropharmacology 12:1081–1086, 1973.
34. VonVoigtlander PF, Losey EG, Triezenberg HJ: Increased sensitivity to dopaminergic agents after chronic neuroleptic treatment. J Pharmacol Exp Therapeut 193:88–94.
35. Weiner WJ, Klawans HL Jr: Lingual, facial, bucal movements in the elderly. I. Pathophysiology and treatment. J Am Geriatr Soc 21:314–317, 1973.
36. Wolfarth S, Dulska E, Lacki M: Comparison of the effects of the intranigral injections of cholinomimetics with systemic injections of the dopamine receptor stimulating and blocking agents in the rabbit. Neuropharmacology 13:867–875, 1974.

The Role of Peptides and Amino Acids as Neurotransmitters, pages 151–160
©1981 Alan R. Liss, Inc., 150 Fifth Avenue, New York, NY 10011

Neuroactive Amino Acids: Effects on C1300 Neuroblastoma Cells

Vernon J. Perez, John C. Eatwell, K. John Morrow, and Robert L. Casady

It has been well established that neuroexcitatory amino acids and their analogs produce neuronal degeneration in discreet areas of the CNS in a variety of animal species [Arees and Mayer, 1970; Lamperti and Blaha, 1976; Olney, 1969; Olney et al, 1973; Olney and Sharpe, 1969; Snapir et al, 1971]. Olney has referred to those amino acids as "excitotoxins" [Olney et al, 1974]. Neuroinhibitory amino acids are without measurable adverse effects on CNS neurons. Neuronal degeneration produced by the excitotoxins is generally confined to the circumventricular organs (CVO), namely, the arcuate nucleus of the hypothalamus (ARH), area postrema (AP), and subfornical organ (SFO) [Olney 1978]. The excitotoxin, glutamate (GLU), accumulates in the ARH, but not in the immediately adjacent ventromedial nucleus of the hypothalamus (VMH) or more dorsally in the medial nucleus of the thalamus (NMT) after subcutaneous administration of the amino acid [Perez and Olney, 1972; Perez et al, 1976; Perez et al, 1979; Perez et al, 1973]. The necrotic effects of the amino acids are the same, regardless of the route of administration, and are produced at considerably lower doses in the neonate than in the mature organism. If the excitotoxin is administered intracerebrally in microinjection, thereby by-passing blood–brain barriers, neuronal degeneration occurs at the site of injection.

The physical characteristics of the neurodegenerative process have been well defined [Olney, 1971]. However, the mechanisms by which the excitotoxic amino acids produce neuronal degeneration are not clear at this time. It has been suggested that the prolonged elevation of GLU, and presumably of other excitotoxins, initiates and maintains a state of depolarization of CVO neurons which ultimately leads to death of the neurons [Olney, 1978; Olney et al, 1975]. This hypothesis has been questioned by Seil et al [1978], among others, on the basis of their failure to measure in vitro any neurotoxic effects of GLU on cultures of infant mouse cerebellar explants or any blocking of the neurotoxic and, presumably, neuroexcitatory effects of kainic acid (KA) by the inhibitory amino acid, γ-aminobutyric acid (GABA) in those explants.

The major purpose of experiments described here was to define the neurotoxic effects of excitotoxic amino acids on neurons in vitro and to compare them with those measured in vivo. The studies were based on the assumption that neurons grown in culture might provide a comparatively less complex biological system than that in vivo and, therefore, might be more amenable to the elucidation and subsequent study of at least some of the mechanisms by which neurotoxic amino acids produce neuronal degeneration.

METHODS

C1300 Azaguanine-resistant mouse neuroblastoma cells (N2AG) used in all experiments were generously provided by Dr. Tom Shows at the Roswell Park Memorial Institute, Buffalo, New York.

Amino acids which were used included the neuroexcitatory L-gluatamate (GLU), L-aspartate (ASP), DL-homocysteic acid (H-CYS) and KA, GABA, and L-alanine (ALA) which are inhibitory amino acids, and L-leucine (LEU) which is not neuroactive. H-CYS was purchased from Cal Biochem–Behring Corporation (La Jolla). GLU was purchased locally as "Accent." The other amino acids were bought from Sigma Chemicals (St. Louis).

In vitro. The stock supply of cells was cultured in Falcon plastic flasks in 10 ml of Dulbecco's modified Eagle's medium containing 10% fetal calf serum. Flasks were kept in an incubator at 37°C, and the medium was changed every other day. For individual experiments the medium was removed and the cells were rinsed with 5 ml Hanks' solution. Cells were dissociated from the flask with 4 ml Versene solution as a single cell suspension. After the suspension was centrifuged in a clinical centrifuge, the Versene was poured off and the cells were resuspended in 3 ml of Hanks' solution. The cells were diluted and pipetted into 60-mm disposable Falcon petri dishes. Three dishes containing 10^3 cells each were used for each concentration of amino acid. Immediately after adding the cells to the dishes, predetermined amounts of amino acids, singly or in combination, were added to the dishes which were then incubated at 37°C for 5 to 7 days. At this time, the medium was poured off, the plates were rinsed with Hanks' solution, and the cells were stained with 1% cresyl violet. The cloning efficiency at each amino acid concentration was calculated by comparing numbers of clones in experimental and control dishes.

In vivo. Five-day-old Cox Swiss albino mice (Laboratory Supply Co., Indianapolis) born in our labs were used in these experiments. Male and female pups were taken randomly from five litters and groups of five were injected once subcutaneously with 0.9% saline, GLU (5.9 mM/kg of body weight), GLU (5.9 mM/kg) plus GABA (4.9, 9.7 or 19.4 mM/kg), GLU (5.9 mM/kg) plus NaCl

(19.4 mM/kg), or GABA (19.4 mM/kg). Concentrations of solutions were prepared so that volumes injected did not exceed 0.1 ml. Immediately after injection, pups were put in a Thelco Model 2 incubator at 32°C for 3 hr. Mice were then anesthetized with pentobarbital sodium and the brains were fixed by transcardial perfusion with 10% Formalin–8.5% sucrose solution [Clark and Clark, 1972]. Brains were removed and stored in the formalin–sucrose solution at 2°C for at least 2 weeks before cutting. Embedding was done in paraffin (Paraplast), sections 8 μm thick were cut through the ARH, and staining was done with cresyl violet as described by Clark and Clark [1972].

RESULTS

In vitro. Results of the experiments with the neuroactive amino acids and with LEU, singly or in combinations, are illustrated in Figures 1–3. Each point on the curves, expressed as percentage of clones in control dishes, represents the mean number of clones identified in three experimental dishes. It is evident in Figure 1 that the neurotoxic amino acids, GLU and H-CYS, inhibited the cloning efficiency of the neuroblastoma cells, with H-CYS showing a greater degree of toxicity than GLU. KA had no detrimental effects on the cells. Variation around each data point (mean for three dishes) ranged from 1.36 (H-CYS, 300 μg/ml) to a maximum of 28.7% (GLU, 200 μg/ml).

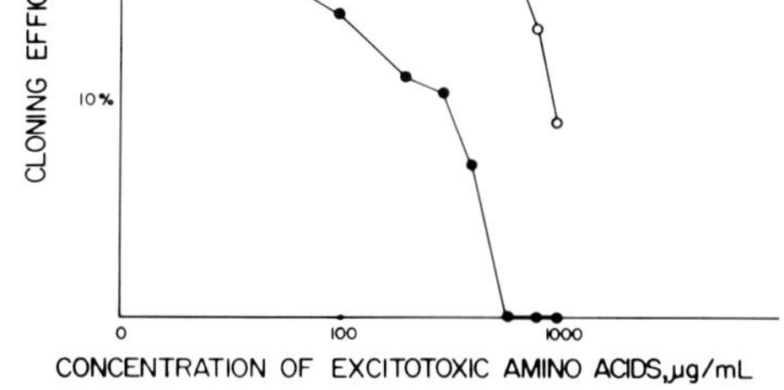

Fig. 1. Cloning efficiency of C1300 neuroblastoma cells in culture medium to which was added the excitotoxins, kainic acid (KA), glutamic acid (GLU), or homocysteic acid (H-CYS). Each point represents the mean number of clones in three culture dishes, expressed as percentage of control. Variation ranged from 1.4 (H-CYS, 300 μg/ml) to 28.7% (GLU, 200 μg/ml).

The effects of LEU, and of the neuroinhibitory amino acids, GABA and ALA, are shown in Figure 2. ALA had no marked adverse effects on the neuroblastoma cells, reducing cloning efficiency to about 80% of control at 800 μg/ml of culture medium. LEU reduced the cloning efficiency to 60% of control at the highest concentration used (1000 μg/ml). GABA tended to increase the cloning efficiency. Variation ranged from 3.01 (LEU, 200 μm/ml) to 49.53% (GABA, 1000 μg/ml).

Addition of the neuroinhibitory amino acids, GABA or ALA, to the incubation medium containing the neurotoxin, GLU, reduced the neurotoxic effects of GLU on the neuroblastoma cells (Fig. 3). It can be seen in Figure 1 that cloning efficiency was reduced to approximately 10% that of control cells by the addition of GLU (1000 μg/ml) to the medium. However, in the presence of GABA (750 μg/ml), GLU failed to reduce cloning efficiency below 50% of control. Furthermore, with the GABA concentration increased to 1000–1500 μg/ml of culture medium, cloning efficiency was not reduced below 80% even though GLU was at a concentration of 1000 μg/ml. Cloning efficiency was not reduced below 78% when ALA (1100 μg/ml) was added to the medium which contained GLU (500 μg/ml). Variation around each of the data points was less than that with the addition of excitatory (Fig. 1) or inhibitory (Fig. 2) amino acids, the maximum being 20.9% for GLU (1000 μg/ml) plus ALA (400 μg/ml).

Neurite complexity was a striking characteristic of the neuroblastoma cultures under control conditions. Addition of GABA or ALA failed to modify this

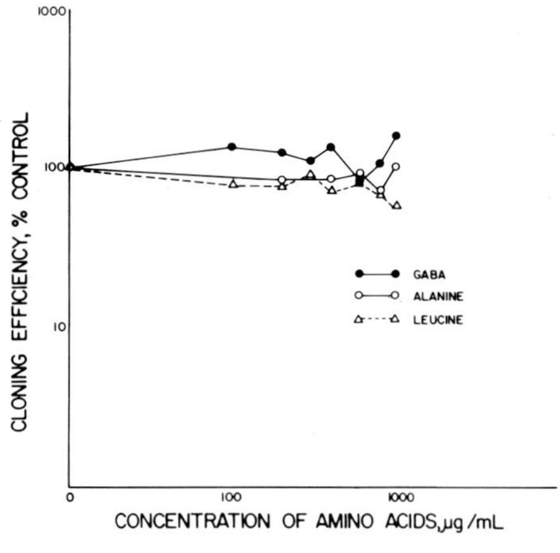

Fig. 2. Cloning efficiency of C1300 neuroblastoma cells in culture medium to which was added leucine (LEU) or the neuroinhibitory amino acids, alanine (ALA), or γ-aminobutyric acid (GABA). Each point represents the mean number of clones in three culture dishes, expressed as percentage of control. Variation ranged from 3.0 (LEU, 200 μg/ml) to 49.5% (GABA, 1000 μg/ml).

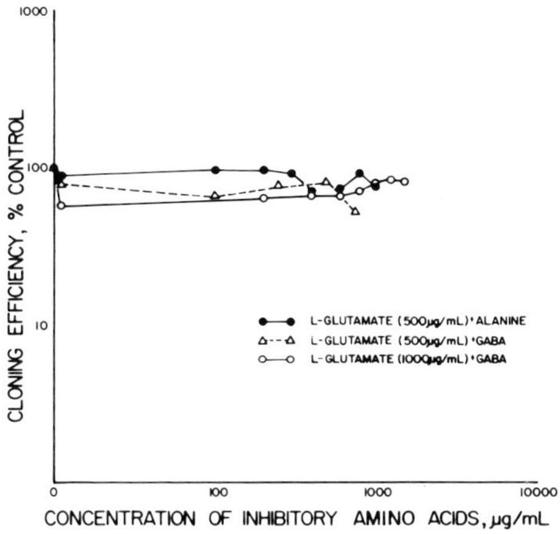

Fig. 3. Cloning efficiency of C1300 neuroblastoma cells in culture medium containing 500 or 1000 μg of GLU/ml of medium, to which was added the neuroinhibitory amino acids, γ-aminobutyric acid (GABA) or alanine (ALA). Each point represents the mean number of clones in three culture dishes, expressed as percentage of control. Variation ranged from 1.1 (GLU, 500 μg/ml; ALA, 600 μg/ml) to 20.9% (GLU, 500 μg/ml; ALA, 400 μg/ml).

appreciably. However, the effects of the excitotoxins, GLU and H-CYS, on the physical characteristics of the clones were pronounced in that the number of clones was severely reduced as were the numbers of cells within a clone and neurite formation for individual cells. At high concentrations of GLU or H-CYS, cell death was prominent and the few detectable cells presented greatly enlarged and grossly distorted profiles. These adverse effects on the clones produced by GLU tended to be limited when GABA or ALA was added to the medium.

In vivo. We described here the histologies for an untreated mouse (Fig. 4A), for an infant killed 3 hr after treatment with GLU (5.9 mM/kg) plus GABA (19.4 mM/kg) (Fig. 4B), and for another injected with GLU (5.9 mM/kg) plus NaCl (19.4 mM/kg) (Fig. 4C). The process of neuronal degeneration, one of the most characteristic features of which is the appearance of dark-staining nuclei [Olney, 1971], is clearly identified 3 hr after injections of GLU (Fig. 4B,C). These dark-staining nuclei are absent in the ARH of the untreated animal (Fig. 4A). It will be noted in Figure 4B that GABA, given simultaneously with GLU at more than three times the molar dose of GLU, failed to inhibit the cytotoxicity produced by GLU. In fact, there was a tendency for the lesion produced by the combined injection of GLU and GABA to be more severe than that produced by GLU plus NaCl. Similarly, NaCl, given with GLU at a dose equimolar to that for GABA (19.4 mM/kg) had no apparent effect on cytotoxic effects of GLU (Fig. 4C).

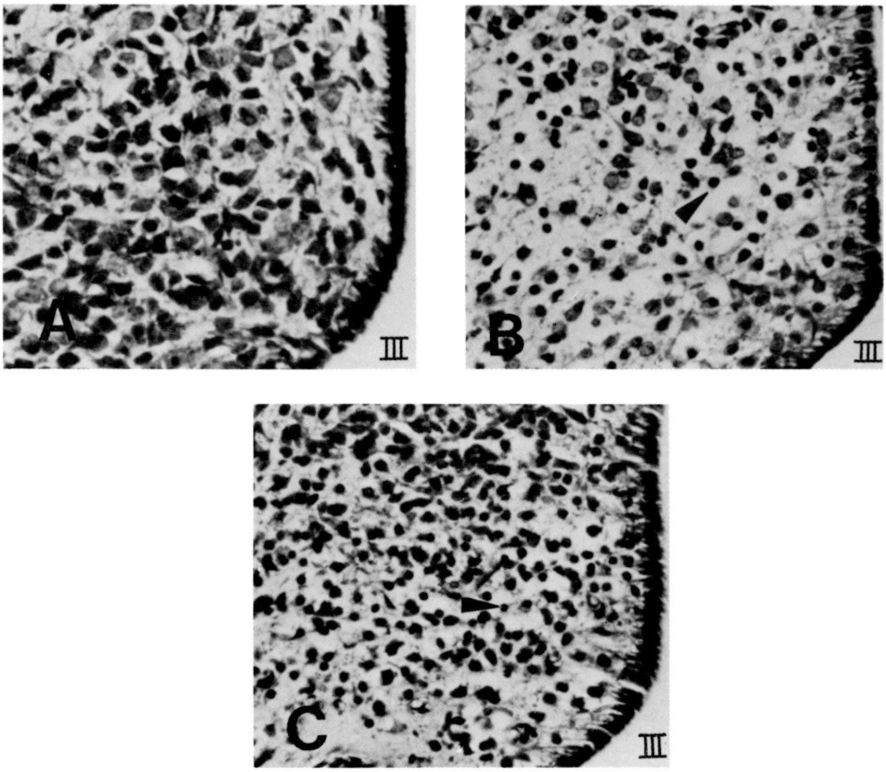

Fig. 4. Light micrographs of the arcuate nucleus of the hypothalamus (ARH) and part of the third ventricle in a 5-day-old untreated control mouse (A), one treated with GLU, 5.9 mM/kg body weight plus GABA, 19.4 mM/kg (B), or GLU, 5.9 mM/kg plus NaCl, 19.4 mM/kg (C). Animals were killed 3 hr after treatment for preparation of paraffin sections and staining with cresyl violet. Dark-staining nuclei in the ARH in animals treated with GLU (B and C) are indicative of neuronal degeneration. GABA failed to block cytotoxicity produced by GLU (B) (Magnification × 200). III is third ventricle.

DISCUSSION

The neurotoxic amino acids, GLU and H-CYS, profoundly inhibited the cloning efficiency of C1300 neuroblastoma cells in culture. Furthermore, neurotoxicity produced by H-CYS was greater than that by GLU. GABA and ALA, two inhibitory amino acids, and LEU, had no severe detrimental effects on the cultured neurons. These results correspond well with those described in vivo by Olney [1978] and Olney et al [1975]. The data suggest, if only tentatively, that cultured neuroblastoma cells may serve as a model system with which to study the mechanisms by which excitotoxic amino acids produce neuronal degeneration.

KA, which is among the most potent excitotoxins, had no detrimental effects on cloning efficiency of the C1300 cells, even at concentrations as high as 1 mg/ml of culture medium. On the other hand, Seil et al [1978] showed that KA produced marked degeneration of Purkinje cells and intracerebellar nucleus cells in cultured infant mouse cerebellar explants. Precise reasons for the difference between those data and those reported here cannot be given at this time but two considerations must be made. First, C1300 cells are tumor cells and may not be strictly comparable to those in the ARH or other circumventricular organs in vivo, or to the Purkinje and related cells in the cultured cerebellar explants used by Seil et al [1978]. Second, the C1300 neuroblastoma cells in our studies were the only cells in the culture medium. However, the cerebellar explants resembled more closely than our cultures conditions found in vivo in that they contained glial cells and neurons. There are structural and functional interactions between neurons and glia in vivo [Gainer et al, 1977; Globus et al, 1973; Hyden and Lange, 1962; Lasek et al, 1977; Monard et al, 1973; Rieske et al, 1975]. KA toxicity may be mediated by the glial cells in this interaction. The mechanisms by which this might occur are unclear at this time. However, it is clear from our data that KA had no cytotoxic effects on C1300 cells in cultures where glial cells were absent, but was shown to be toxic to neuronal populations in cultured cerebellar tissues in which glia were present.

Results of in vivo experiments reported here suggest that the cytotoxic effects of neuroexcitatory amino acids, at least of GLU, on C1300 cells could be partially blocked by adding GABA or ALA to the culture medium. However, in the work reported by Seil et al [1978], GABA failed to block the cytotoxicity of KA in cultures of mouse cerebellum. This suggests that the cytotoxicity of KA may be mediated through mechanisms different from those for GLU and other endogenous excitatory amino acids [McGeer and McGeer, 1978; McGeer et al, 1978]. Neurotoxicity produced by GLU might involve prolonged depolarization of the affected cells as originally suggested by Olney [1978] and Olney et al, [1975] and might be blocked, if only partially, by the hyperpolarizing effects of GABA as the data from our in vitro studies could suggest. This interpretation receives indirect support if one considers that there is little evidence to suggest that glia respond to stimulation by excitatory amino acids [Varon and Somjen, 1979] and that they are resistant to degenerative changes in the presence of these amino acids [Olney, 1971]. However, the cytotoxic effects of KA may depend less, if at all, on prolonged depolarization of neuronal membranes and more on the mediating effects of glial cells in the neuronal–glial interaction—an interaction which may be unresponsive to the effects of GABA. If neurotoxic actions of KA are not associated with electrophysiological changes in cell membranes, then the hyperpolarizing actions of neuroinhibitory amino acids might not be expected to be effective in blocking KA cytotoxicity.

Our failure to demonstrate any interference by GABA on GLU cytotoxicity in vivo might be accounted for by the failure to use an effective GABA:GLU dose ratio. That is, had GABA been administered at a higher dose in relation to GLU, some degree of blocking action might have been observed. Another possibility is that in vivo, GABA failed to cross the blood–brain barrier to enter the ARH with GLU [Perez and Olney, 1972] and, therefore, could not have been expected to have interferred with the toxic effects of GLU in the ARH. In vitro, there was no blood–brain barrier, and GABA and ALA were free to interact with GLU. However, the virtual absence of a blood–brain barrier in the ARH would argue against this suggestion.

Data from experiments reported here suggest that the in vitro system in which C1300 neuroblastoma cells are grown in cultures may serve as a useful model for the study of cytotoxic mechanisms of neuroexcitatory amino acids. However, limitations of the system must be kept in mind. As already noted, C1300 neuroblastoma cells are tumor cells and may not be strictly comparable to those found in vivo. Furthermore, neurons grown in cultures in the absence of glial cells may not represent neuronal populations in vivo in which neuronal-glial interactions are prominent. That there are differences between cultured cells and those in vivo is suggested by the data reported by Passonneau et al [1977] showing marked differences between activities of enzymes in the GABA shunt in mouse brain, on the one hand, and in cultured C1300 neuroblastoma and C6 astrocytoma cell lines, on the other.

SUMMARY

The mechanisms by which neuroexcitatory amino acids destroy neurons in discreet areas in the CNS are undefined. Studies reported here were done to measure the response of neurons to excitatory and inhibitory amino acids in vitro. It was postulated that the cultured cells provide a comparatively simpler biological system in which to study the degenerative process than that in vivo and might be a valid model for the study of cell death. C1300 Azaguanine-resistant mouse neuroblastoma cells, and glutamate (GLU), homocysteic acid (H-CYS), kainic acid (KA), γ-aminobutyric acid (GABA), alanine (ALA), and leucine (LEU) were used. Five milliliters of Dulbecco's modified Eagles medium containing 10% fetal calf serum were put in individual dishes to which were added 1000 cells. Cultures were incubated for 5 to 7 days at 37°C, and cells were stained with cresyl violet. GLU and H-CYS inhibited growth and cloning efficiency of the cells, reduced numbers of cells in individual clones, and virtually eliminated growth of neurites from individual cells in a clone. The more potent an amino acid as an excitant in vivo, the greater the toxicity in vitro. KA, among the most potent excitotoxins, had no adverse effects on the C1300 cells. ALA and LEU, at relatively high concentrations, reduced cloning efficiency to 60–80% of control. GABA had no consistent effect on cloning of the cells. When GABA or ALA was added to the medium containing GLU, cytotoxicity produced by GLU was greatly reduced. These data are inconsistent with those obtained in vivo showing GABA to have no effect on the cytotoxicity produced by GLU.

Additional studies are in order in which metabolic events in cultured neurons and glial cells are measured during the degenerative process produced by excitotoxins. It is also imperative that the role of glial cells in the neurodegenerative process be studied. The structural and functional interactions between neurons and glial cells in the CNS have been well established and may be essential for maintaining the viability of neuronal populations. Our data suggest that glial cells may be involved in the neurodegenerative process associated with KA cytotoxicity. The in vitro system is not without its limitations in the study of excitotoxins. However, the relative simplicity of the cellular population lends itself well to this area of research.

REFERENCES

1. Arees EA, Mayer J: Monosodium glutamate-induced brain lesions: Electron microscopic examination. Science 170:549–550, 1970.
2. Clark G, Clark MP: "A Primer in Neurological Staining Procedures." Springfield: C.C. Thomas, 1972.
3. Gainer H, Tasaki I, Lasek RJ: Evidence for the glia–neuron protein transfer hypothesis from intracellular perfusion studies of squid axons. J Cell Biol 74:524–530, 1977.
4. Globus A, Lux HD, Schubert P: Transfer of amino acids between neuroglia cells and neurons in the leech ganglion. Exp Neurol 40:104–113, 1973.
5. Hyden H, Lange PW: A kinetic study of the neuron-glia relationship. J Cell Biol 13:233–237, 1962.
6. Lamperti A, Blaha G: The effects of neonatally administered monosodium glutamate on the reproductive system of adult hamsters. Biol Reprod 14:362–369, 1976.
7. Lasek RJ, Gainer H, Barker JL: Cell-to-cell transfer of glial proteins to the squid giant axon. The glia-neuron protein transfer hypothesis. J Cell Biol 74:501–523, 1977.
8. McGeer EG, McGeer PL: Some factors influencing the neurotoxicity of intrastriatal injections of kainic acid. Neurochem Res 3:501–517, 1978.
9. McGeer PL, McGeer EG, Hattori T: Kainic acid as a tool in neurobiology. In McGeer EG, Olney JW, and McGeer PL (eds): "Kainic Acid as a Tool in Neurobiology." New York: Raven, 1978, pp 123–138.
10. Monard D, Solomon F, Kentsch M, Gysin R: Glia-induced morphological differentiation in neuroblastoma cells. Proc Nat Acad Sci 70:1894–1897, 1973.
11. Olney JW: Brain lesions, obesity and other disturbances in mice treated with monosodium glutamate. Science 164:719–721, 1969.
12. Olney JW: Glutamate-induced neuronal necrosis in the infant mouse hypothalamus. An electron microscopic study. J Neuropathol Exp Neurol 30:75–90, 1971.
13. Olney JW: Neurotoxicity of excitatory amino acids. In McGeer EG, Olney JW, and McGeer PL (eds): "Kainic Acid as a Tool in Neurobiology." New York: Raven, 1978, pp 95–121.
14. Olney JW, Ho OL, Rhee V, deGubareff T: Neurotoxic effects of glutamate. N Eng J Med 289:1374–1375, 1973.
15. Olney JW, Sharpe LG: Brain lesions in an infant rhesus monkey treated with monosodium glutamate. Science 166:386–388, 1969.
16. Olney JW, Sharpe LG, deGubareff T: Excitotoxic amino acids. Neurosci Abstr 1:371, 1975.
17. Passonneau JV, Lust WD, Crites SK: Studies on the GABAergic system in astrocytoma and neuroblastoma cells in culture. Neurochem Res 2:605–617, 1977.

18. Perez VJ, Olney JW: Accumulation of glutamic acid in the arcuate nucleus of the hypothalamus of the infant mouse following subcutaneous administration of monosodium glutamate. J Neurochem 19:1777–1782, 1972.
19. Perez VJ, Olney JW, Frolichstein CF, Martin JF, Cannon WO: Regional CNS uptake of neurotoxic and nontoxic amino acids by the infant mouse brain. Biochem Pharmacol 25:1415–1419, 1976.
20. Perez VJ, Olney JW, Martin JF, Cannon WO: Minimal tissue concentrations of glutamate required to produce necrosis of hypothalamic neurons in newborn mice. Biol Neonate 35:17–22, 1979.
21. Perez VJ, Olney JW, Robin SJ: Glutamate accumulation in infant mouse hypothalamus: Influence of temperature. Brain Res 59:181–189, 1973.
22. Rieske E, Schubert P, Kreutzberg GW: Transfer of radioactive material between electrically coupled neurons of the leech central nervous system. Brain Res 84:365–382, 1975.
23. Seil FJ, Woodward WR, Blank NK, Leiman AL: Evidence against chronic depolarization as a mechanism of kainic acid toxicity in mouse cerebellar cultures. Brain Res 159:431–435, 1978.
24. Snapir N, Robinzon B, Perek M: Brain damage in male domestic fowl treated with monosodium glutamate. Poultry Sci 50:1511–1514, 1971.
25. Varon SS, Somjen GC: Neuron–glia interactions. Neurosci Res Prog Bull 17:1979.

The Role of Peptides and Amino Acids as Neurotransmitters, pages 161–169
©1981 Alan R. Liss, Inc., 150 Fifth Avenue, New York, NY 10011

Stimulation of Hypothalamic Arcuate Nucleus Inhibits Locus Coeruleus Unit Activity: Evidence for Endorphin Mediation

Howard K. Strahlendorf, Jean C. Strahlendorf, and Charles D. Barnes

Since the original isolation of endogenous opiate compounds by Hughes [1975], two independent endogenous opioid systems have been delineated: a β-endorphin system emanating from a group of cells within or slightly lateral to the arcuate nucleus of the hypothalamus, which sends long axons to midbrain and limbic structures [Bloom et al, 1978], and a short axon enkephalin system with multiple cell groups throughout the spinal cord, brain stem, and encephalon [Watson et al, 1977]. Opiate receptors on locus coeruleus (LC) neurons [Atweh and Kuhar, 1977] presumably mediate the effects of endorphins released from concentrated stores in nerve terminals surrounding LC neurons [Simantov et al, 1977]. A source of these terminals is believed to be the β-endorphin-containing cell bodies of the hypothalamus [Bloom et al, 1978; Watson et al, 1980]. Congruently, noradrenergic cells of the LC are inhibited by systemic and iontophoretic administration of opiates [Bird and Kuhar, 1977; Korf et al, 1978] and enkephalins [Cedarbaum and Aghajanian, 1978].

Experimental data [Guillemin et al, 1977; Korf, 1976; Ward et al, 1976] suggest an anatomical framework for reciprocal modulatory influences between the LC and hypothalamus and, thus, indirectly the adenohypophysis. Certain areas of the hypothalamus have been proposed to be under a noradrenergic influence orignating from the LC [Cross, 1973; Ward et al, 1976; Palkovits et al, 1980], whereas, the LC may be modulated by peptide-containing neurons emanating from the arcuate nucleus of the hypothalamus. The present investigation examines the effect of this endorphin projection on LC neurons.

METHODS

Chloralose-anesthetized (60 mg/kg, ip), immobilized (Flaxedil, 2 mg/kg/hr), artificially ventilated cats were used in experiments where drugs were given intravenously. Male Sprague–Dawley rats anesthetized with chloral hydrate (400 mg/kg ip) were used in iontophoretic studies. Spontaneously active single units encountered within the region of the LC were recorded with stainless-steel microelectrodes (cats) or conventional 1-, 5-, or 7-barrel micropipettes (rats and cats). Action potentials were amplified and displayed on an oscilloscope. Signals were also fed into a window discriminator, the output of which generated a ratemeter record as well as a dot display on a second storage oscilloscope which was controlled by a raster stepper. LC neurons were initially identified at the time of recording by various electrophysiologic criteria, ie, rhythmic discharge and slow firing rate (0.5–3.0 Hz) [Bird and Kuhar, 1977], response to nociceptive input applied to digits of the forepaw (bursting followed by inhibition) [Cedarbaum and Aghajanian, 1978], and failure to respond to mandible manipulation. Brief trains (three to five shocks) of constant current pulses were delivered to the region of the ipsilateral arcuate nucleus of the hypothalamus through concentric bipolar electrodes. A standard iontophoretic circuit with current balancing was used to apply drugs to LC neurons directly.

Drug solutions for systemic administration were prepared in physiologic saline, and all doses refer to the salt form of the compound. Control injections were equivolumes of saline. Cats received drugs via a cannulated femoral vein, while drugs were given to rats by way of a tail vein catheter. Drugs used for iontophoresis were prepared in distilled water and consisted of the following: morphine sulfate 0.05 M, pH 4.5; naloxone hydrochloride 0.05 M, pH 4.0; piperoxane hydrochloride 0.05 M, pH 4.0; and monosodium glutamate 0.1 M, pH 8.0. Histologic verification of all stimulating and recording electrode placements was done at the end of each experiment utilizing the Prussian blue reaction or by localizing a fast green dye spot.

RESULTS

Electrical activation of the arcuate region of the hypothalamus (25–500 μA, 50–100 Hz) produced profound inhibition of spontaneously active locus coeruleus cells in cats and rats. Arcuate nucleus stimulation elicited total suppression of cell firing lasting 652 msec (range of 200–1500 msec, n = 30 cells) which was succeeded by an epoch of partial inhibition lasting an additional 295 msec (range up to approximately 2100 msec). The inhibition frequently displayed an immediate onset but occasionally (35% of total cells) was preceded by a brief discharge of typically one or two spikes within 50 msec of arcuate stimulation. In cats, maximum inhibition of LC neurons occurred when the stimulating electrode was located 1 mm lateral of the midline, 1 mm above the base of the brain, and in an area between 11 and 13 mm anterior to stereotaxic zero. Placements outside these boundaries failed to elicit characteristic inhibition. Arcuate-induced inhibition in rats occasionally was less profound, lacking in duration and magnitude.

The inhibitory influence of the arcuate nucleus on LC was examined by calculating the percentage of inhibition of cell firing for a period of time after arcuate nucleus stimulation inclusive of the period of total inhibition. This time period was defined in controls by the onset of arcuate nucleus stimulation and by a time at which cell activity appeared to be no longer affected. Naloxone reduced the degree of inhibition from 67 (± 8 SEM) to 27% (±9 SEM, seven cells). Occasionally, naloxone increased the spontaneous firing rate of nonconditioned LC units (n = 4 cells) (Fig. 1).

In cats, morphine sulfate (0.25, 0.5, 1.0 mg/kg cumulative, iv) produced a dose-dependent, naloxone reversible, decrease in spontaneous coerulear cell activity (Fig. 2), which was preceded at the lowest dose by a transient period (3–4 min) of excitation. Morphine also elicited a dose-related augmentation of arcuate-derived suppression of coerulear cell activity. Inhibition of cell activity increased in a graded fashion from 37% before morphine (saline) to 49, 70, and 75% after cumulative doses of 0.25, 0.5, and 1.0 mg/kg of this opiate, respectively. Accordingly, naloxone antagonized this activity by effectively reversing the inhibition to beyond premorphine levels (Fig. 2). In contrast to naloxone, arcuate-induced inhibition was augmented by the low dose of morphine during the period of enhanced cell firing, negating the possibility that naloxone blockade of arcuate suppresion was due to a nonspecific increase in unit activity. Since enhancement of arcuate actions by morphine occurred in the presence of a higher firing rate of LC neurons, the possibility that the similar effect observed after the higher doses of morphine might be due to a general reduction in the cell discharge rate elicited by the opiate-induced decrease in the rate of rise of Excitatory Postsynaptic Potentials (EPSPs) is excluded.

Iontophoretically applied naloxone in rats and cats has demonstrated essentially similar findings. Naloxone ejected with currents of 10–20 nA for a period of 3–10 min reversibly attenuated arcuate-elicited suppression of LC cell firing (Fig. 3). These same cells were also slowed by iontophoretically applied morphine; this action was sensitive to blockade by concomitant application of naloxone (data not shown).

To establish the stereospecificity of the depressant response to opiates, dextrorphan was administered in cumulative doses of 0.25, 0.50, and 1.0 mg/kg. Dextrorphan failed to depress the spontaneous LC unit discharge rate and appeared to antagonize arcuate-induced inhibition of coerulear units (Fig. 2). In all instances, control injections were ineffective in altering either spontaneous cell firing or arcuate-elicited suppression of single unit discharges.

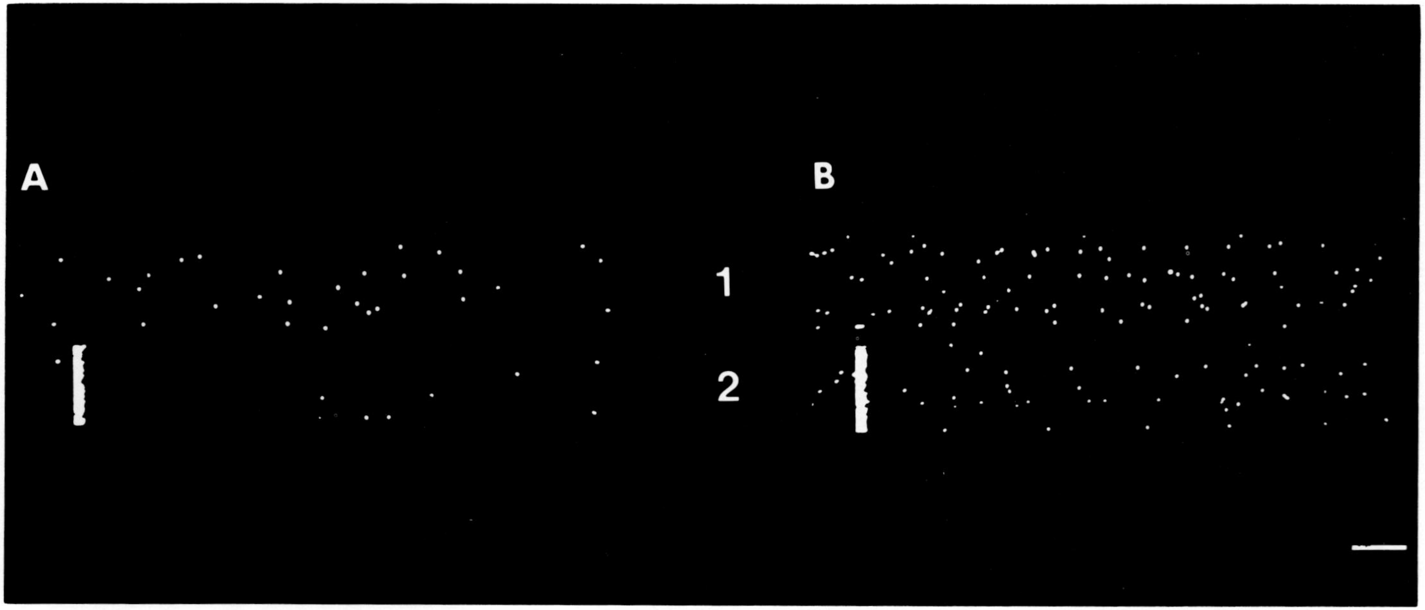

Fig. 1. Naloxone antagonism of arcuate-induced inhibition of a locus coeruleus cell. Each dot represents an action potential which met the criteria of the window discriminator. Each group is 32 sweeps. (A) Control. (B) Naloxone, 5.0 mg/kg. (1) Spontaneous activity. (2) Arcuate stimulation, three pulses, 200 μA, 0.1 msec, 100 Hz (calibration, 200 msec).

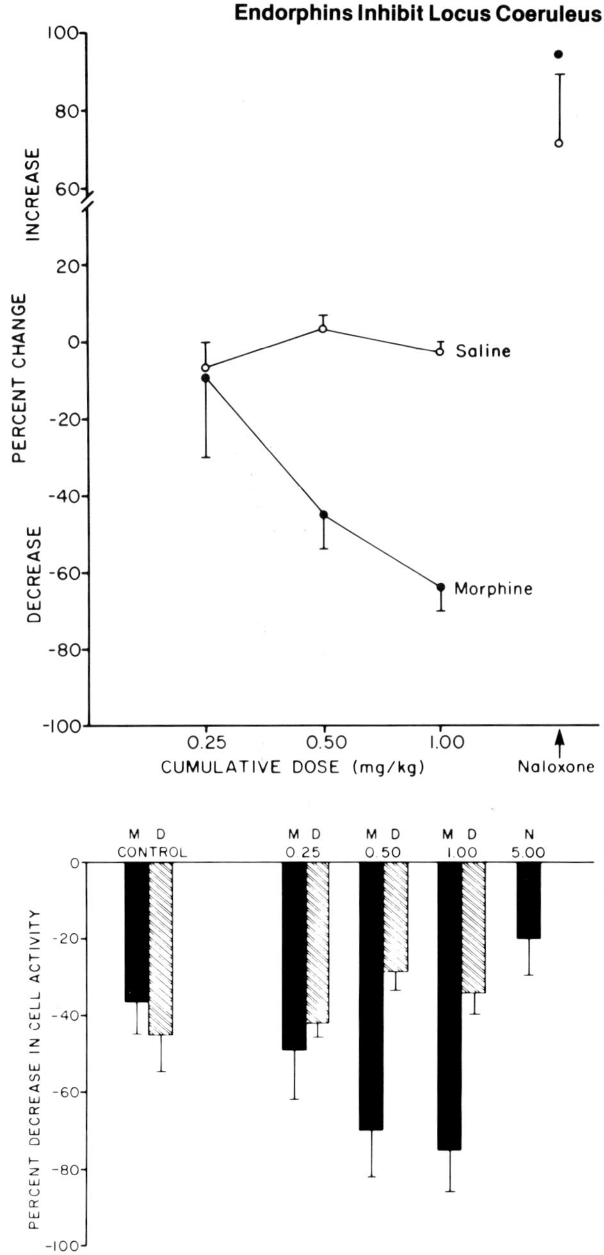

Fig. 2. Effects of morphine sulfate (M), dextrorphan tartrate (D), and naloxone HCl (N) on spontaneous activity and arcuate-induced inhibition of locus coeruleus neurons. Top, (○), equivolume saline injection. (●), morphine sulfate. Each point is the mean percentage of change from predrug activity levels ± SEM of three or four experiments. Bottom, each bar is the average percentage of change ± SEM from spontaneous activity occurring during a 2-sec sweep repeated 16 times. Control, saline injection equivolume to largest drug injection, n = 4 cells; M, n = 6 cells; D, n = 6 cells; N, n = 4 cells. Doses are cumulative milligrams per kilogram, iv.

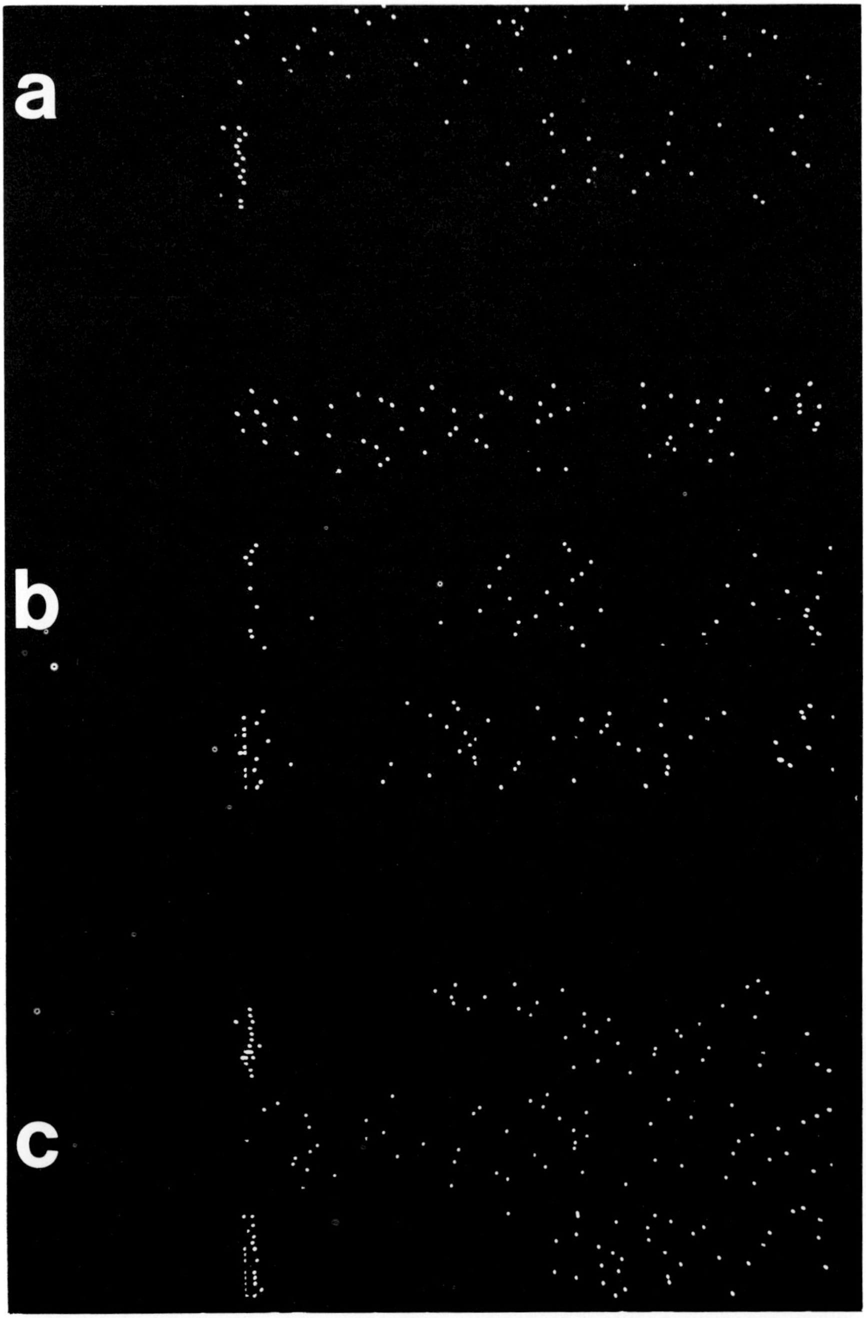

Fig. 3. Antagonism by iontophoretically applied naloxone of spontaneous coerulear unit activity and inhibition from arcuate stimulation in rats. (A) Predrug; top, spontaneous activity; bottom arcuate stimulation, 50 μA, 100 Hz. (B) Naloxone, 1.5 NA. Responses at 3 min (top, spontaneous firing), 5 min (middle, arcuate stimulation), and 10 min (bottom, arcuate stimulation) after initiation of application. (C) Recovery, top and bottom 5 and 10 min after cessation of naloxone current with arcuate stimulation, middle spontaneous activity 7 min after turning off naloxone. Each raster group is 16 sweeps (calibration, 200 msec).

DISCUSSION

The present findings electrophysiologically impart functional significance to the neuronal network emanating from cell bodies within the arcuate nucleus, which project to, interalia, the LC [Bloom et al, 1978]. That arcuate nucleus stimulation inhibited LC unit firing, as presently reported, supports the concept that the latter nucleus is influenced by endogenously released opiate peptides. In accord with this hypothesis, microiontophoretically applied endorphins have been shown to exert predominantly depressant effects on neuronal activity in various CNS sites [Bradley et al, 1976; Frederickson and Norris, 1976]. Intravenous as well as iontophoretic naloxone markedly antagonized arcuate-induced suppression of LC unit activity and increased spontaneous cell firing rate lending additional credence to the notion that inhibition of coerulear units occurred in response to activation of opiate receptors by endogenously released β-endorphin. Significantly, these data imply that neurons of the LC may be tonically influenced by the arcuate projection under the present experimental situations.

The LC projects rostrally to the hypothalamus [Palkovits et al, 1980]. Conceivably, stimulation of the arcuate nucleus in the present experiments could inhibit LC unit firing by means of an axon reflex, ie, autoinhibition. This appears untenable in that in the majority of instances, arcuate stimulation produced profound, naloxone-antagonizable inhibition in the absence of any initial discharge, ie, antidromic activation.

Present dose–response data reveal that LC cell firing rate was substantially reduced by intravenous morphine, and, on several occasions, the lowest dose of morphine transiently increased coerulear cell firing rate. Although no immediate explanation for this effect is apparent, systemic morphine may release LC cells from tonic arcuate inhibition by displacing β-endorphin from opiate receptors. Since morphine possesses less intrinsic agonistic activity than β-endorphin [Miller and Cuatrecasas, 1979], an increase in cell activity would result until sufficient receptors were occupied by morphine.

Additionally, morphine augmented the period of arcuate-induced inhibition in a dose-related fashion, suggestive of additive effects on LC opiate receptors. That inhibition of LC unit activity by arcuate stimulation and exogenous morphine reflected activity at a common set of receptors is substantiated by the fact that naloxone completely reversed both effects. It is recognized that naloxone antagonism could also occur at sites other than arcuate–LC synapses and thus indirectly affect LC neurons. Current microiontophoretic experiments suggest that naloxone exerts a primary blockade at arcuate–LC synapses.

That the LC is profoundly inhibited by the arcuate–endorphin system implies a reciprocal governing feedback system between these two areas which could function to maintain homeostasis during stress. Release of β-endorphin concomitantly with ACTH and in equimolar amounts suggests that peptide neurotransmitters could inform the LC of the degree of hypothalamic response to stress.

SUMMARY

We have investigated the effect of arcuate nucleus stimulation on locus coeruleus unit activity. Extracellular LC unit activity was recorded in anesthetized cats and rats. Brief conditioning trains (25–300 μA, three shocks, 100 Hz) delivered to the arcuate nucleus produced profound inhibition (average 1000 msec, n = 30) of spontaneous coerulear cell discharges. Intravenously administered naloxone (2.5–5.0 mg/kg) or iontophoretically applied naloxone eliminated arcuate-elicited inhibition. Additionally, naloxone often increased spontaneous cell activity. In contrast to naloxone, systemically administered morphine inhibited spontaneous LC unit activity in a dose-related fashion and augmented the hypothalamic-derived inhibition. Naloxone, 5 mg/kg, reversed morphine activity to beyond predrug levels. To characterize the stereospecificity of the response, dextrorphan was administered. This inactive enantiomer failed to mimic morphine actions on LC cells. The results suggest that LC may be influenced by the β-endorphin system originating from the arcuate nucleus.

ACKNOWLEDGMENTS

We wish to thank Endo Laboratories for the gift of naloxone hydrochloride and Hoffman–La Roche Incorporated for the gift of dextrorphan tartrate. This work was supported in part by the Tarbox Parkinson's Disease Institute of TTUSM and NIH Grant HL-7289.

REFERENCES

1. Atweh SF, Kuhar MJ: Autoradiographic localization of opiate receptors in rat brain: II. The brain stem. Brain Res 129:1–12, 1977.
2. Bird SJ, Kuhar MJ: Iontophoretic application of opiates to the locus coeruleus. Brain Res 122:523–533, 1977.
3. Bloom FE, Rossier J, Battenberg EL, Bayon A, French E, Henriksen SJ, Siggins GR, Segal D, Browne R, Ling N, Guillemin R: Beta-endorphine: Cellular localization, electrophysiological and behavioral effects. Advan Biochem Psychopharmacol 18:89–109, 1978.
4. Bradley PD, Briggs I, Gayton RJ, Lambert LA: Effects of microiontophoretically applied methionine enkephalin on single neurones in rat brain stem. Nature (London) 261:425–426, 1976.
5. Cedarbaum JM, Aghajanian GK: Activation of locus coeruleus neurons by peripheral stimuli: Modulation by a collateral inhibitory mechanism. Life Sci 23:1383–1392, 1978.
6. Cross BA: Unit responses in the hypothalamus. In Ganong WF and Martini L (eds): "Frontiers in Neuroendocrinology." New York: Oxford University Press, 1973, pp 133–171.
7. Frederickson RCA, Norris FH: Enkephalin-induced depression of single neurones in brain areas with opiate receptors: Antagonism by naloxone. Science 194:440–442, 1976.
8. Guillemin R, Vargo T, Rossier J, Minick S, Long N, Rivier C, Bloom F: Beta-endorphin and adrenocorticotropin are secreted concomitantly by the pituitary gland. Science 197:1367–1369, 1977.
9. Hughes J: Isolation of an endogenous compound from the brain with pharmacological properties similar to morphine. Brain Res 88:295–308, 1975.

10. Korf J: Locus coeruleus, noradrenaline metabolism and stress. In Usdin E (ed): "Catecholamines and Stress." Oxford: Pergamon Press, 1976, pp 105–111.
11. Korf J, Bunney BS, Aghajanian GK: Noradrenergic neurons: Morphine inhibition of spontaneous activity. Eur J Pharmacol 25:165–169, 1974.
12. Miller RJ, Cuatrecasas P: Neurobiology and neuropharmacology of the enkephalins. Advan Biochem Psychopharmacol 20:187–225, 1979.
13. Palkovits M, Zaborszky L, Feminger A, Mezey E, Fekete MIK, Herman JP, Kanyicska B, Szabo D: Noradrenergic innervation of the rat hypothalamus: Experimental biochemical and electron microscopic studies. Brain Res 191:161–171, 1980.
14. Pert CB, Kuhar MJ, Snyer SH: Autoradiographic localization of the opiate receptor in rat brain. Life Sci 16:1849–1854, 1975.
15. Simantov R, Kuhar MJ, Uhl GR, Snyder SH: Opioid peptide enkephalin: Immunohistochemical mapping in rat central nervous system. Proc Nat Acad Sci 74:2167–2171, 1977.
16. Ward DG, Grizzle WE, Gann DS: Inhibitory and facilitatory areas of the rostral pons mediating ACTH release in cat. Endocrinology 99:1220–1228, 1976.
17. Watson SJ, Akil H, Sullivan S, Barchas JD: Immunocytochemical localization of methionine enkephalin: Preliminary observations. Life Sci 21:733–738, 1977.
18. Watson SJ, Richard III CW, Ciaranello RD, Barchas JD: Interaction of opiate peptide and noradrenalin systems: Light microscopic studies. Peptides 1:23–30, 1980.

The Role of Peptides and Amino Acids as Neurotransmitters, pages 171–179
©1981 Alan R. Liss, Inc., 150 Fifth Avenue, New York, NY 10011

Taurine-Induced Thalamocortical Dissociative Effects in an Experimental Model of Epilepsy

J.B. Lombardini and Tamas L. Frigyesi

The enigma of taurine, its high concentration and uneven distribution in the mammalian central nervous system (CNS) [Lombardini, 1976] has recently prompted numerous studies. Taurine has been claimed to possess antiepileptic properties in both animal modes [Van Gelder, 1972; Fariello et al, 1975; Durelli et al, 1976; Mutani et al, 1978; Roches et al, 1979] and human epilepsies [Bergamini et al, 1974; Mutani et al, 1978; Takahashi and Nakane, 1978; Barbeau et al, 1976]. It has been speculated that these effects of taurine might be related to its depressant effects on synaptic transmission in many areas of the central nervous system [Curtis and Tebecis, 1972; Haas and Hösli, 1973; Krnjević and Puil, 1976]. However, other reports suggest that taurine does not have anti-epileptic properties [Wada et al, 1975; Joseph and Emson, 1976] or, on the contrary, is an epileptogenic agent [Frigyesi and Lombardini, 1979].

The present study was originally set out to reinvestigate the results of studies that claimed that taurine levels in the brains of epileptic animals were decreased and that administration of taurine prevented experimental seizures [Mutani et al, 1977, 1978; Van Gelder, 1978]. Although our investigation has failed to corroborate these studies, it has provided new information about mechanisms of supraspinal seizures.

METHODS

Adult cats were initially anesthetized with intravenous sodium thiamylal (30 mg/kg), then paralyzed with succinylcholine chloride and artificially ventilated. While under the barbiturate anesthesia, the neuraxis was cut between C_1 and the medulla. The entire left hemisphere and the cerebellar surface were then exposed by extensive craniectomy. All branches of the trigeminal nerve and exposed tissues were liberally, and repeatedly, infiltrated with 2% Xylocain [Frigyesi et al, 1974; Frigyesi and Machek, 1976]. Body temperature was maintained at 37 ± 0.5° by a servo-controlled heating coil wrapped around the body and governed by a rectal thermistor.

Within 3–5 hr after the short-acting barbiturate was administered, an acute epileptogenic focus was produced in the lateral 1/5 of the left anterior sigmoid gyrus by sodium penicillin G (20,000 IU) powder. Surface activities were recorded from the left motor cortex (MC) by an Ag–AgCl ball electrode embedded in the penicillin powder. Focal potentials were recorded from the left thalamic ventrolateral nucleus (VL) and the right cerebellar Purkinje (P) layer by sharpened, Epoxy-insulated tungsten wires.

VL neurons were identified by their monosynaptic responsiveness to stimulation of the right brachium conjunctivum (BC) and antidromic responsiveness to stimulation of the corona radiata, immediately below the silver-ball electrode. P cells were identified antidromically by stimulating the right nucleus interpositus.

The cortical surface, and thalamic and cerebellar focal potentials were amplified by RC-coupled preamplifiers (at 1 KHz, 0.21-sec time constants) and displayed on an AC-coupled channel of an oscilloscope. Data were recorded on continuous films.

Attempts were made to record "on-line" from neuronal pools in the motor cortical penicillin focus and VL. Previous intracellular observations have revealed that one type of VL neurons exhibits both antidromic and monosynaptic potentials when fine stimulating electrodes are used to activate a restricted number of axons in the corona radiata, adjacent to the motor cortex [Frigyesi and Machek, 1976]. This topographical organization of the reciprocal thalamocortical paths permitted us to record from relatively closely couple thalamic and motor cortical neuronal pools. In addition, fine concentric stimulating electrodes (barrel-tip separation: 0.02–0.03 mm) were inserted into the right BC; the experimental position of the silver-ball electrode was selected at the left motor cortical region where BC-evoked potentials exhibited identical electrographic properties to both polarities of stimulation. When such a BC–motor cortical relationship was established, a microelectrode was inserted into VL; an intra-VL site was sought where BC elicited focal potentials with identical electrographic properties to both polarities of stimulation.

Taurine was administered iv in incrementing doses of 25–50 mg/kg, up to a total of 500 mg/kg (pH 7.3, buffered with potassium phosphate, 0.04 M).

The taurine content of the spinal cord and brain regions was determined on centrifuged perchloric acid extracts (2 vol of 2% perchloric acid) on a Beckman 121 automatic amino acid analyzer.

RESULTS

The Rank order of taurine levels in 13 regions of the adult feline CNS is shown in Figure 1. Taurine was unevenly distributed in the CNS; the highest concentrations were observed in the cerebellum and the lowest concentrations were in the spinal cord.

Figure 2 demonstrates that there is a progressive decrease in the concentration gradient from the cerebellum, through the cortex and related corpus callosus, as well and the caudate nucleus, toward the phylogenetically older structures of the brain stem and the spinal cord.

The variations in concentrations of taurine were studied in 8 out of these 13 regions of the CNS. Figure 3 demonstrates recordings from three cats that were killed under identical conditions, as it was judged by physiological criteria (left side of Fig. 3) of status epilepticus; cat B exhibited no change, cat C increased, and cat A decreased concentrations of taurine (right side of Fig. 3). However, a review of all data in 54 cats has revealed that cerebral taurine exhibited random variations during epileptic status.

The electrophysiological effects of iv administration of taurine were studied in the motor cortex, related thalamic nuclei, and Purkinje layer of the cerebellum

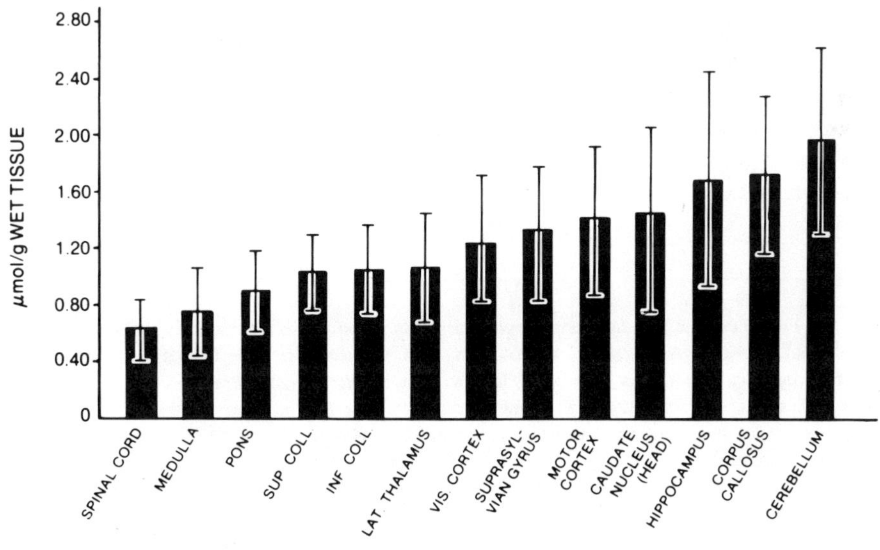

Fig. 1. Rank order of taurine levels in 13 regions of the adult feline central nervous system. Data are presented as means ± standard deviations from 30 cats.

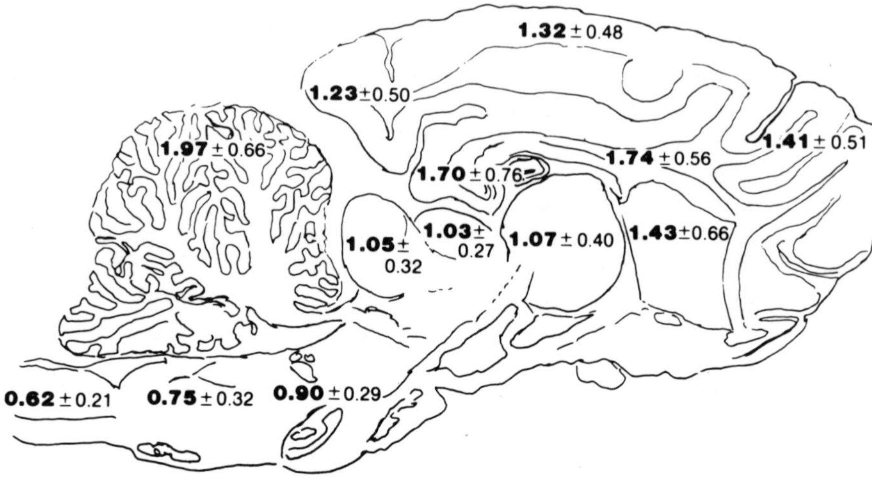

Fig. 2. Representation of the feline central nervous system with quantity of taurine (μmoles per gram of tissue) illustrated in the various regions. Data are presented as means ± standard deviations from 30 cats.

(Figs. 4–6). Figure 4 demonstrates that taurine increased the number of VL neurons that rhythmically discharged during the epileptic status (B). In this animal, detectable taurine-induced motor cortical effects were not observed during the initial 23 min following the iv administration of taurine (B). Subsequently, however, the magnitude of cortical spikes decreased without concomitant decreases in the frequency of VL discharges (C).

Figure 5 illustrates a more extreme example of taurine-induced dissociations in corticothalamic functional relations during an approximate 3-sec period of

---

Fig. 3. Prominently different taurine levels from brains which exhibit overtly similar states of paroxysmal activities. A, B, and C are data from three consecutive experiments. Left hand side: upper traces, focal recordings from left VL; lower traces: from the left motor cortex. Recordings from each animal are continuous. The time marker in C refers to all traces in this figure. In each animal one fully developed ictus and the beginning of a subsequent ictus are shown; the postictal depressions which separate the two icti in each experiment exhibit comparable durations (approximately 10 sec). Recordings in these animals were taken 4 min prior to removal of the left motor cortex where the penicillin focus was situated. Levels of taurine (μmole per gram of wet tissue) from various parts of each brain are tabulated on the right hand side of the figure. IPSI and CONTRA denote laterality of sampling with respect to the penicillin focus. (From [Frigyesi and Lombardini, 1978] with permission.)

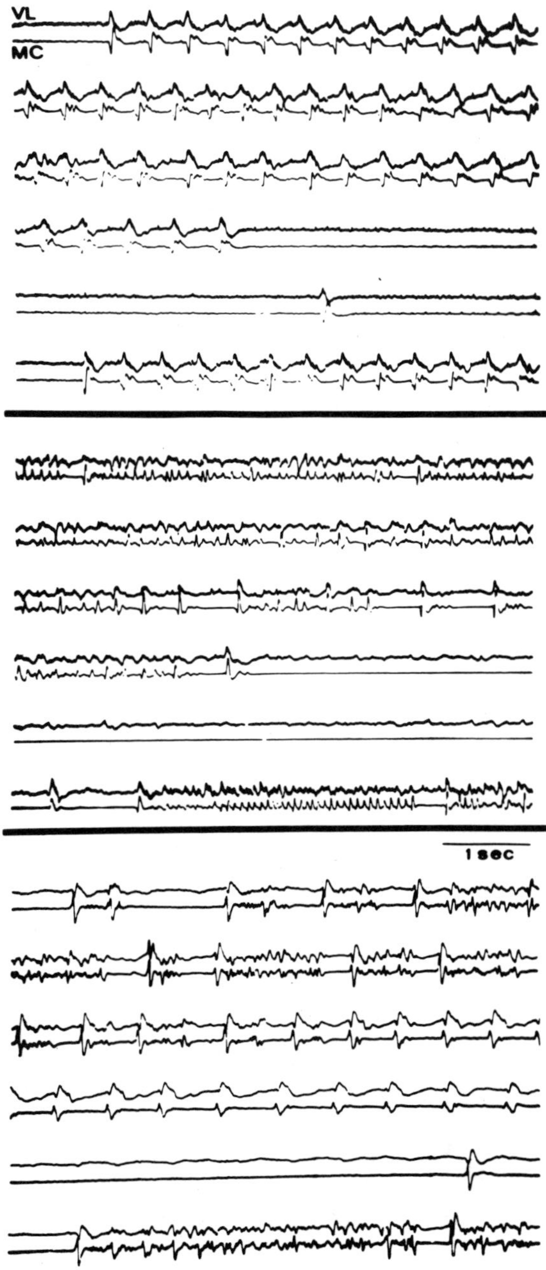

A

	IPSI	CONTRA
MOTOR CORTEX	0.67	0.81
ASSOCIATION C	0.70	0.73
VISUAL CORTEX	0.77	0.76
HIPPOCAMPUS	1.11	0.84
THALAMUS	0.75	0.56
CAUDATE NUCL	0.78	0.96
CEREBELLUM	1.20	1.26
MEDULLA	0.48	0.55

B

	IPSI	CONTRA
MOTOR CORTEX	1.96	1.54
ASSOCIATION C	1.58	1.38
VISUAL CORTEX	1.38	1.33
HIPPOCAMPUS	2.56	2.12
THALAMUS	1.02	1.03
CAUDATE NUCL	2.27	2.30
CEREBELLUM	2.50	2.46
MEDULLA	0.73	0.81

C

	IPSI	CONTRA
MOTOR CORTEX	2.89	3.15
ASSOCIATION C	2.96	3.37
VISUAL CORTEX	2.91	3.19
HIPPOCAMPUS	3.46	3.90
THALAMUS	2.30	2.08
CAUDATE NUCL	3.58	3.98
CEREBELLUM	3.33	3.53
MEDULLA	1.09	1.34

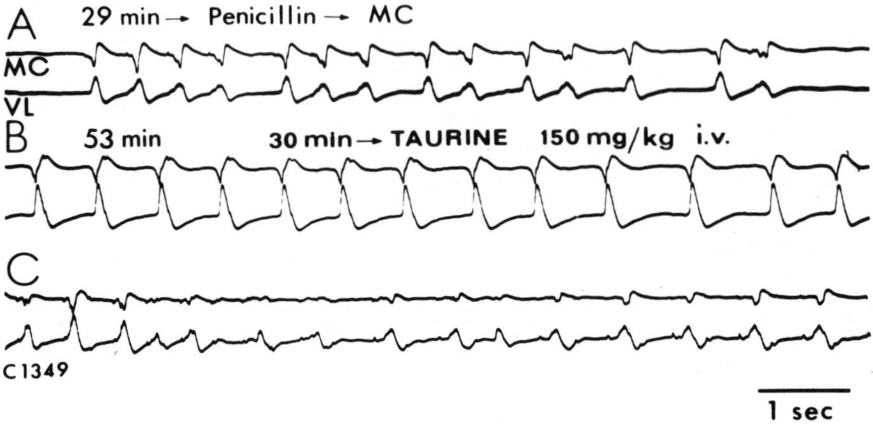

Fig. 4. Disparate effects of taurine on motor cortex and ventrolateral tracings on epileptogenesis induced by a penicillin focus in the motor cortex. Taurine (150 mg/kg) was administered iv 30 min after the penicillin focus was initiated. Tracing are shown at 29 min (A, prior to taurine administration); and 53 min (B, 23 min after taurine administration).

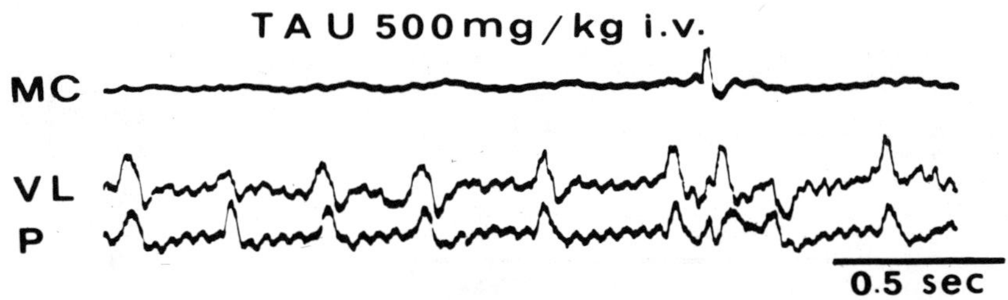

Fig. 5. Effects of iv taurine (500 mg/kg) on the time relationships between the motor cortex, ventrolateral thalamus, and Purkinje potentials. Penicillin was placed on the motor cortex.

epileptic status; only one cortical discharge was noticeable pari passo the persisting discharges. In addition to these effects, taurine induced *spatial* (corticothalamic) and prominent *temporal* (in the sequential order of ictal firings in the motor cortex, VL, and cerebellar Purkinge layers) dissociations (Fig. 6). Following iv administration of taurine, VL and Purkinje discharges (in B) relative to the cortical spike occurred 180 msec earlier than during the pretaurine period (in A).

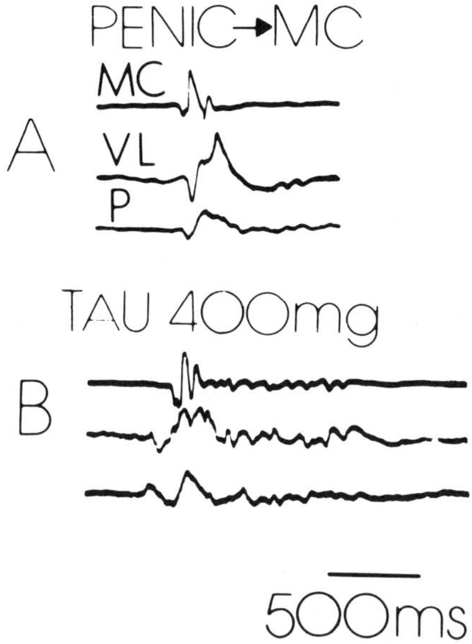

Fig. 6. Effects of iv taurine (400 mg/kg) on the time relationships between the motor cortex, ventrolateral thalamus, and Purkinje potentials. Penicillin was placed on the motor cortex.

DISCUSSION

Taurine in the cat is an essential amino acid [Sturman and Hayes, 1980]. The uneven, and as this study shows, orderly distribution of taurine in the feline brain suggests that taurine has an active role in higher nervous activities. The present study supports the contention that taurine plays a role in the feline epileptogenesis [Fariello, 1975]. However, this study has revealed that taurine is an epileptogenic agent when administered iv and that cerebral regional concentrations of taurine cannot be correlated with seizure states induced by a penicillin focus in the motor cortex [Frigyesi and Lombardini, 1978].

Several investigators recording from only the cortex showed that taurine decreased the recordable number of spikes; from this it was inferred that taurine is antiepileptic [Mutani et al, 1978; Roches et al, 1979; Van Gelder, 1972, 1978]. This study indicates that this inference is more apparent than real. In this study, we chose to concomitantly investigate motor cortical and VL and Purkinje cell activities because these structures are functionally, intimately related to the integration of sensorimotor activities in general [Frigyesi, 1976; Dray, 1980] and in experimental modes of epilepsy [Frigyesi, 1976] in particular. Concomitant electrophysiological recordings from the cortex and thalamus have revealed that during periods of attenuation [Mutani et al, 1978; Roches et al, 1979] or even abolition [Tsukamoto et al, 1978] of cortical interictal and ictal spikes by iv taurine, thalamic seizure activities were exacerbated [Lombardini and Frigyesi, 1978; Frigyesi, and Lombardini, 1979]. These studies indicated that only neocortical recordings, to assess taurine effects in the CNS, are insufficient; coincident recordings from the thalamus and cerebellum have revealed that taurine induced dissociative effects on penicillin epileptogenesis [Lombardini and Frigyesi, 1979; Frigyesi and Lombardini, 1979]. Consequently, prudence requires that additional cerebral areas, like hypocampus and caudate nucleus that were demonstrated here to exhibit high content of taurine, be included in the determination of taurine-induced epileptogenic events during penicillin and other types of experimental epilepsy.

SUMMARY

The distribution of taurine in the feline CNS was demonstrated to be unevenly distributed: The highest concentrations were observed in the cerebellum and the lowest were observed in the spinal cord. The taurine content was determined in 16 brain regions, including the penicillin focus during status epilepticus. Under this condition of status epilepticus, cerebral taurine exhibited random variations. Administration of iv taurine increased the number of VL neurons that discharged during the epileptic status, while no effects were initially observed in the motor cortex. Taurine induced dissociations in cortiocothalamic relationships during epileptic status.

ACKNOWLEDGMENTS

This work was suported in part by NIH Research Grant 1-RO1-11406 (J.B. Lombardini). Ms. Evangeline V. Medina and Mr. Harvey O. Olney are thanked for their technical assistance.

REFERENCES

1. Barbeau A, Tsukada A, Naohide I: Neuropharmacologic and behavioral effects of taurine. In Huxtable R and Barbeau A (eds): "Taurine." New York: Raven Press, 1976, pp 253–266.
2. Bergamini L, Mutani R, Delsedime M, Durelli L: First clinical experience on the antiepileptic action of taurine. Eur Neurol 11:261, 1974.
3. Curtis DR, Tebecis AK: Bicuculline and thalamic inhibition. Exp Brain Res 16:210–218, 1972.
4. Dray A: The physiology and pharmacology of mammalian basal ganglia. Prog Neurobiol 14:221–335, 1980.
5. Durelli L, Mutani R, Delsedime M, Quattrocolo G, Buffa C, Mazzarino M, Fumero S: Electroencephalographic and biochemical study of the antiepileptic action of taurine administered by cortical superfusion. Exp Neurol 52:30–39, 1976.
6. Fariello RG, Lloyd KG, Hornykievicz O: Cortical and sub-cortical projected foci in cats: Inhibitory action of taurine. Neurology 25:1077–1083, 1975.
7. Frigyesi TL: Parkinsonism and epilepsy. In Messiha FS and Kenny AD (eds): "Parkinson's

Disease, Neurophysiological, Clinical, and Related Aspects." New York: Plenum Press, 1976, pp 63–102.

8. Frigyesi TL, Lombardini JB: Lack of correlation between taurine levels in 16 brain regions and paroxysmal discharges in the thalamocortical circuit. Neurosci Let 7:213–217, 1978.
9. Frigyesi TL, Lombardini JB: Augmentation of thalamo-motor cortico-cerebellar epileptogenesis by taurine and its antagonism by diphenylhydantoin. Life Sci 24:1251–1260, 1979.
10. Frigyesi TL, Machek J: Basal ganglia-diencephalon synaptic relations in the cat. An intracellular study of dorsal thalamic neurons during capsular and basal ganglia stimulation. Brain Res 20:201–217, 1976.
11. Frigyesi TL, Yahr MD, Schwartz R: Electrophysiological analysis of the effects of L-Dopa on the reciprocal corticothalamic projections in the squirrel monkey. I. Iatrogenic coma and disinhibition of medial thalamic neurons. J Neural Transm 35:151–173, 1974.
12. Haas HL, Hösli: The depression of brain stem neurons by taurine and its interaction with strychnine and bicuculline. Brain Res 52:399–402, 1973.
13. Joseph MH, Emson PC: Taurine and cobalt induced epilepsy in the rat: A biochemical and electrocorticographic study. J Neurochem 27:1495–1501, 1976.
14. Krnjević K, Puil E: Electrophysiological studies on actions of taurine. In Huxtable R and Barbeau A (eds): "Taurine." New York: Raven Press, 1976, pp 179–189.
15. Lombardini JB: Regional and subcellular studies on taurine in the rat central nervous system. In Huxtable R and Barbeau A (eds): "Taurine." New York: Raven Press, 1976, pp 311–326.
16. Lombardini JB, Frigyesi TL: Cerebro-cerebellar functional relationships in an experimental epilepsy (penicillin focus). Anat Rec 190:463–464, 1978.
17. Lombardini JB, Frigyesi TL: Thalamocortical dissociative effects induced by a penicillin focus in the motor cortex. Anat Rec 193:168, 1979.
18. Mutani R, Bergamini L, Durelli L: Taurine in experimental and human epilepsy. In Barbeau A and Huxtable R (eds): "Taurine and Neurological Disorders." New York: Raven Press, pp 359–374, 1978.
19. Mutani R, Durelli L, Mazzarino M, Valentini C, Monaco F, Fumero S, Mondini A: Longitutidinal changes of brain amino acid content occurring before, during and after epileptic activity. Brain Res 122:513–521, 1977.
20. Roches JC, Zumstein HR, Fassler A, Scollo-Lavizzari G, Hösli L: Effects of taurine, glycine and GABA on convulsions produced by strychnine in the rabbit. Eur Neurol 18:26–32, 1979.
21. Sturman JA, Hayes KC: The biology of taurine in nutrition and development. Advan Nutr Res 3:231–299, 1980.
22. Takahashi R, Nakane Y: Clinical trial of taurine in epilepsy. In Barbeau A and Huxtable R (eds): "Taurine and Neurological Disorders." New York: Raven Press 1978, pp 375–386.
23. Tsukamoto Y, Frigyesi TL, Roberts LA, Lombardini JB: The effects of taurine and diphenylHydantoin (DPH) on penicillin epileptogenesis (cat). Fed. Proc 37:347, 1978.
24. Van Gelder NM: Antagonism by taurine of cobalt induced epilepsy in cat and mouse. Brain Res 47:157–165, 1972.
25. Van Gelder NM: Glutamic acid and epilepsy: The action of taurine. In Barbeau A and Huxtable R (eds): "Taurine and Neurological Disorders." New York: Raven Press, 1978, pp 387–402.
26. Wada JA, Osawa T, Wake A, Corcoran ME: Effects of taurine on kindled amygdaloid seizures in rats, cats, and photosensitive baboons. Epilepsia 16:229–234, 1975.

The Role of Peptides and Amino Acids as Neurotransmitters, pages 181–189
© 1981 Alan R. Liss, Inc., 150 Fifth Avenue, New York, NY 10011

Amino Acid and Monoamine Inhibition of Sacral Parasympathetic Preganglionic Neurons

Thomas D. Thomson and Donald N. Franz

Histochemical studies of central monoaminergic pathways indicate that sacral parasympathetic preganglionic neurons (PPGNs) receive dense innervation by bulbosacral 5-hydroxytryptamine (5-HT) neurons and lesser innervation by bulbosacral norepinephrine (NE) neurons [Dahlström and Fuxe, 1965]. Since the possible functions of these descending pathways remain unknown, the present study was originated to assess their roles by testing the effects of monoamine precursors on central transmission through PPGNs. In addition, the study furnished evidence for an unusual spinal inhibitory system utilizing glycine and γ-aminobutyric acid (GABA).

METHODS

Adult cats were made spinal under ether anesthesia, paralyzed with gallamine triethiodide, and maintained on artificial respiration. Blood supply to the head was occluded. Carotid blood pressure, end-tidal CO_2, and temperature were monitored and maintained optimal. Drugs were administered through an indwelling catheter in a cephalic vein.

Preganglionic parasympathetic nerve activity was recorded with bipolar, silver wire electrodes from the central portion of the left pelvic nerve, isolated by a dorsal approach through the greater sciatic notch and removal of lateral sections of sacral vertebrae and overlying muscle and sacral somatic nerves. In several experiments, the bladder was cannulated through a suprapubic incision of the urethra to record spontaneous and evoked changes in bladder pressure.

Spinal parasympathetic reflexes were evoked by single-pulse stimulation (0.2 msec, 0.2 Hz) of sacral somatic nerves or exposed dorsal roots. Parasympathetic discharges were also evoked by single-pulse intraspinal stimulation (0.1 Hz) of descending excitatory pathways in the dorsolateral funiculus of the exposed spinal cord at T8 with bipolar, tungsten microelectrodes. Several experiments were performed on decerebrate cats in which pelvic nerve activity was evoked by stimulation of medullary vesico-constrictor centers as described by Kuru [1965].

Evoked pelvic nerve discharges were amplified and recorded on an oscilloscope and were analyzed on-line at 5- or 10-min intervals by integrating sums of 32 consecutive responses with a signal-averaging computer (Nicolet 1072). Evoked responses were generally quite stable ($< \pm$ 10%) for 5–6 hr or longer. Drugs were tested only after stability was established during a control period of at least 1 hr.

RESULTS

Preganglionic parasympathetic discharges evoked by stimulation of spinal reflex, intraspinal, or medullary pathways and recorded from the pelvic nerve were quite small, seldom larger than 20 μV, but prolonged (100–150 msec). However, very large discharges could be evoked by stimulation of sacral ventral roots which contain the parasympathetic outflow, suggesting that most of the available PPGNs were not activated by the excitatory pathways, possibly due to some inhibitory process. The failure of low spinal transection or complete lumbosacral dorsal rhizotomy to increase the small sizes of evoked spinal reflex discharges indicated that higher spinal or peripheral afferent activity did not contribute to the apparent inhibition of PPGNs.

Effects of Convulsant Drugs

Since both glycine and GABA are inhibitory transmitters in the spinal cord, the possibility that PPGNs are under the influence of a local tonic inhibition by glycine or GABA was tested by determining the effects of their antagonists. As shown by the intraspinal experiment illustrated in Figure 1A and B, repeated administration of strychnine SO_4 increased the size of intraspinally evoked discharges by only about 50%, but subsequent administration of picrotoxin rapidly increased the discharges to almost 10 times their original size. If the order of the antagonists was reversed (Fig. 1C and D), repeated doses of picrotoxin alone increased the evoked responses only slightly, but subsequent administration of a small dose of strychnine rapidly increased their size fourfold. This pattern was consistent in every experiment; large increases were produced only by the addition of the second drug.

The effect of the strychnine–picrotoxin combination on pelvic nerve discharges was somewhat greater upon intraspinal than spinal reflex pathways, but was independent of the order in which they were administered (Table I). In two experiments on decerebrate cats in which discharges were evoked by stimulation of medullary vesico-constrictor centers, this combination increased their sizes to 416 and 900% of control levels after strychnine or picrotoxin alone had

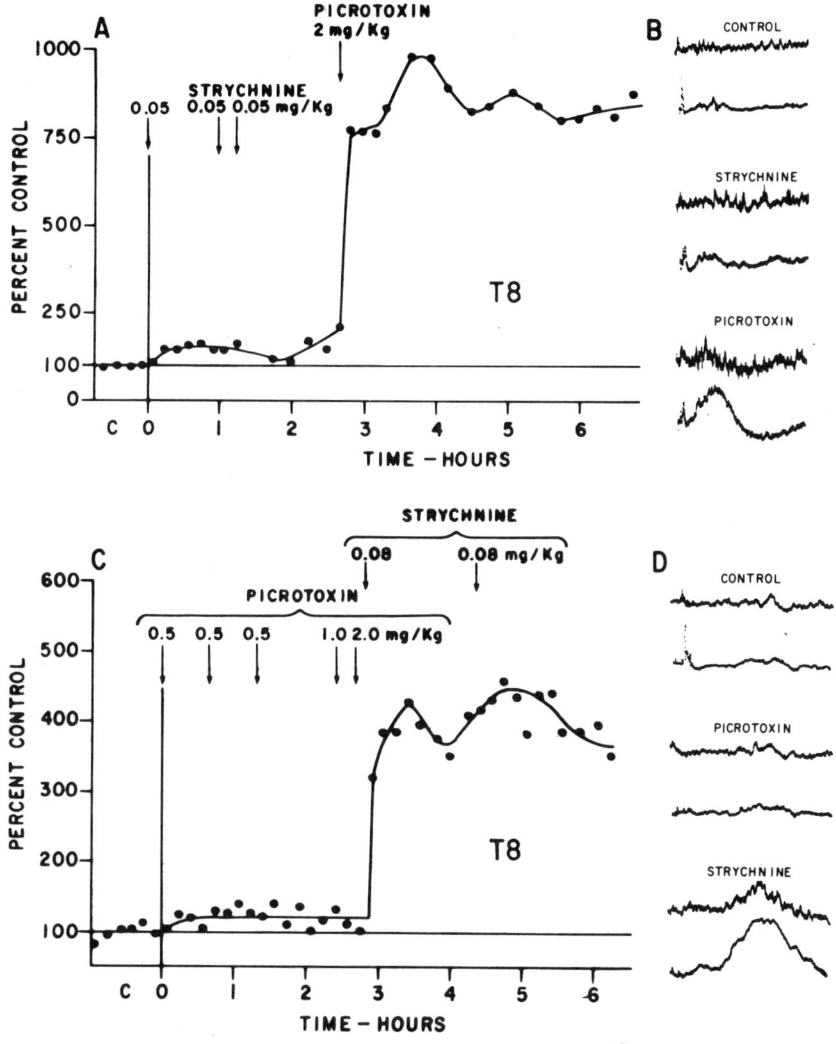

Fig. 1. Effects of strychnine and picrotoxin alone and in combination on sizes of pelvic nerve discharges evoked by stimulation of descending spinal pathways at T8 in two spinal cats. (A) and (B) Time course and representative traces showing increases by strychnine alone and after addition of picrotoxin. (C) and (D) Drugs administered in reverse order. Upper traces in (B) and (D) are single responses; lower traces are computer sums of 32 consecutive responses. Duration of traces is 500 msec in (B) and 250 msec in (D); vertical calibration in (D) represents 20 μV for upper traces in (B) and (D).

TABLE I. Enhancement of Spinal Reflex and Intraspinal Transmission by Convulsant Drugs Alone and in Combination*

Drugs[a]		Spinal reflex			Intraspinal		
(1st)	(2nd)	(1st)	(2nd)	(n)	(1st)	(2nd)	(n)
Picro.	Strych.	105 ± 3	369 ± 107	6	140 ± 20	474 ± 58	12
Strych.	Picro.	108 ± 4	348 ± 120	3	148 ± 15	560 ± 85	9
Bicuc. or Strych.	Both	120 ± 4	497 ± 175	2	110 ± 7	359 ± 57	4
Semicarb.	Strych.	160	583	1	155	590	1

*Values expressed as mean ± SEM % of control values.
[a]Drug doses (mg/kg): picrotoxin (2.0), strychnine (0.1–0.2), bicuculline (1.0), semicarbazide (200).

produced increases of only 26 and 8%, respectively. In general, maximal increases were induced by combinations of 0.2 mg/kg of strychnine SO_4 and 2 mg/kg of picrotoxin. Enhancement of evoked activity either remained at peak levels or stabilized at lower levels for 2–6 hr before declining. In the latter case, additional doses of one or both drugs usually increased the size of evoked responses again. No difference was noted between the degree enhancement of spinal reflexes in high spinal (C1) or low spinal (T10) preparations. In 32 experiments on the three pathways, the combination of strychnine and picrotoxin produced an average increase to 475%; in several experiments the maximum increases exceeded 900%. The latencies of evoked discharges were unchanged, but their durations usually increased by 50–100 msec.

The strychnine–picrotoxin combination also increased spontaneous pelvic nerve activity which was reflected in the appearance of large spontaneous bladder and colon contractions mediated through the intact contralateral pelvic nerve. In addition, the normally small increases in bladder pressure (3–5 cm of H_2O) evoked by brief trains of intraspinal stimulation (20 Hz for 2 sec) were increased markedly to 20–50 cm of H_2O.

Combinations of strychnine with bicuculline (1.0 mg/kg) were as effective as those with picrotoxin in increasing evoked pelvic nerve discharges, but bicuculline alone produced only modest increases (Table I). Likewise, partial depletion of central GABA levels by 200 mg/kg of semicarbazide [Bell and Anderson, 1972] produced only modest increases, but subsequent administration of 0.1 mg/kg of strychnine after 4 hr markedly increased responses evoked by either intraspinal or spinal reflex pathways (Table I).

Effects of Monoamine Precursors

Administration of L-5-HTP (12.5–50 mg/kg) or D,L-5-HTP (25–50 mg/kg) produced dose-dependent depression of pelvic nerve discharges evoked through either spinal reflex or intraspinal pathways in 16 experiments (Fig. 2A). Both

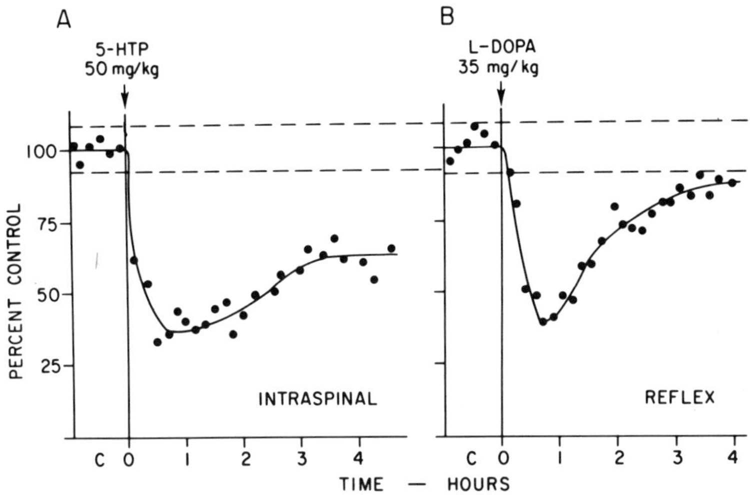

Fig. 2. (A) Time course of depression of intraspinally evoked pelvic nerve discharges by D,L-5-HTP. (B) Time course of depression of spinal parasympathetic reflexes by L-dopa. Dashed lines represent ± 2 SD from mean control (C) values.

pathways were depressed about equally for several hours; the L-isomer was twice as potent as the D,L-racemate. Responses were depressed to about 45 and 25% of control values by 25 and 50 mg/kg of L-5-HTP, respectively. Administration of chlorimipramine (5 mg/kg) alone produced 20–40% depression and further enhanced the depressant effects of 5-HTP. For example, the initial average depression of spinal reflexes to 63% by 17.5 mg/kg of L-5-HTP in four experiments was increased to 36% of control values by chlorimipramine.

L-Dopa (25–35 mg/kg) also depressed transmission through both spinal reflex and intraspinal pathways in 13 experiments to about 60% of control values (Fig. 2B). Pretreatment with parachlorophenylalanine (PCPA; 100 mg/kg, ip for 3 days) did not alter this depressant effect. Protriptyline (5 mg/kg) alone produced 10–20% depression and enhanced that produced by L-dopa to about 45%.

Evoked discharges that had been increased by strychnine–picrotoxin were depressed by 5-HTP or L-dopa to the same degree as discharges that had not been enhanced. The potency of both precursors was enhanced 5–10 times by pretreatment with pargyline (30 mg/kg).

DISCUSSION

The dramatic increases in evoked parasympathetic discharges produced by simultaneous antagonism of both glycine and GABA contrast markedly with the modest increases produced by antagonism or depletion of either inhibitory amino acid alone. The very large increases in size of discharges evoked through either spinal reflex or bulbospinal pathways were consistently produced only by combinations of strychnine, a selective glycine antagonist, with either picrotoxin or bicuculline, both selective GABA antagonists [Davidson, 1976]. Strychnine was equally effective in GABA-depleted cats. The consistency and uniformity of these results indicate that sacral PPGNs are strongly inhibited by a local, tonically active inhibitory mechanism which is mediated by a combination of both glycine and GABA. Blockade or depletion of either inhibitory transmitter alone is inadequate to overcome this inhibition. Since the inhibition does not depend on input from the periphery or higher centers in the brain or spinal cord, it must be generated locally and spontaneously within the caudal spinal cord. To our knowledge, such an inhibitory system in the central nervous system has not been described before.

Our studies provide no information on the synaptic arrangements involved in the glycine–GABA inhibition of PPGNs. However, iontophoretic studies indicate that PPGNs are depressed by both glycine and GABA [de Groat, 1970]. The depressant action of glycine is blocked by strychnine, but that by GABA is not blocked by picrotoxin. Strychnine also antagonizes recurrent inhibition to PPGNs, but apparently at a remote site [de Groat, 1976]. Therefore, whereas strychnine may block glycine receptors on PPGNs as well as elsewhere, picrotoxin and, presumably, bicuculline may block GABA receptors only at a different site.

Since micturition normally requires some deliberate concentration, and even decerebrate but not spinal animals micturate almost normally [Barrington, 1921; de Groat and Ryall, 1969], it seems necessary to postulate the presence of a bulbospinal pathway that disinhibits the tonic glycine–GABA inhibitory system to PPGNs. That this system is distinct from medullary vesico-constrictor centers is suggested by the failure of repetitive stimulation of these centers or their spinal pathways to evoke discharges of sacral antonomic neurons that were comparable in size to those revealed by the convulsant combinations. The origin and characteristics of such a disinhibiting pathway are unknown.

Decentralization of the local tonic glycine–GABA inhibition of PPGNs may account, at least partially, for the persistent loss of bladder function following spinal cord injuries or transection. In most cases, micturition can eventually be evoked through spinal reflex pathways by tactile stimulation of perineal, perianal, or other cutaneous receptors below the spinal lesion [de Groat, 1975; de Groat and Booth, 1980]. This return of functional capacity at the spinal level may reflect a gradual reduction in the local inhibition mediated by glycine and GABA, improvement in the synaptic efficacy of segmental pathways, or both.

The monoamine precursors, 5-HTP and L-dopa, only depressed evoked parasympathetic discharges in both untreated and convulsant-treated cats. These effects appear to be mediated by release of 5-HT or NE from their respective bulbospinal terminals, since the effects were enhanced by inhibiting their reuptake and the potency of the precursors was markedly enhanced by inhibiting monoamine oxidase. Therefore, both monoaminergic pathways appear to depress the excitability of PPGNs. This conclusion is consistent with the iontophoretic studies of Ryall and de Groat [1972]. The depressant effects of 5-HTP on transmission through PPGNs also agree with the results of similar experiments reported by de Groat et al [1979]. In contrast to previous findings on sympathetic preganglionic neurons [Franz et al, 1977] in which depressant effects of L-dopa were prevented and reversed by depletion of 5-HT with PCPA, such treatment did not alter the depressant effects of L-dopa on transmission through PPGNs. Therefore, whereas bulbospinal NE pathways appear to be excitatory to the sympathetic neurons, they appear to be inhibitory to the parasympathetic neurons. The 5-HT pathways appear to inhibit both autonomic neurons.

The present study provides little information on the respective origins or functional roles of the bulbospinal monoaminergic pathways to PPGNs other than that they appear to be inhibitory. The close proximity of vesico-relaxer centers [Kuru, 1965] to the medullary origins of 5-HT neurons suggests that 5-HT might mediate inhibition of PPGNs from these centers. However, the depressant effect of L-dopa is inconsistent with the location of NE neurons in areas of the brainstem that activate sacral parasympathetic neurons upon stimulation [Kuru, 1965, de Groat, 1975]. Both pathways probably function independently of the localized GABA–glycine inhibitory system, since both precursors produced equivalent depression whether this inhibitory system was blocked or not.

The consistent finding that tricyclic antidepressants enhanced the depressant effects of the precursors suggests that their effectiveness in treating nocturnal enuresis is due at least in part to their ability to enhance transmission between the inhibitory monoaminergic pathways and the sacral parasympathetic neurons. Similar actions in higher centers may also contribute. The rapid impairment of reuptake systems and enhancement of monoaminergic transmission by tricyclic antidepressants [Sangdee and Franz, 1979] are not consistent with and cannot account for the slow onset of their antidepressant effects. However, the rapid onset and long duration of their antienuretic effect are perfectly consistent with their ability to block reuptake and enhance central monoaminergic transmission. The onset and duration of increased functional bladder capacity produced by imipramine in conscious cats [Shaffer et al, 1979] are also compatible with central enhancement of monoaminergic inhibition of PPGNs, especially since peripheral actions are largely eliminated as possible mechanisms.

Although the present study focused on tricyclics selective for either 5-HT or NE reuptake instead of imipramine, the standard drug for enuresis, other tricyclic drugs are also effective [Blackwell and Currah, 1973]. Furthermore, our previous study on the acute effects of a series of tricyclics on monoaminergic transmission to sympathetic preganglionic neurons [Sangdee and Franz, 1979] demonstrated that imipramine is about equally effective in enhancing transmission by both NE and 5-HT. The other tricyclics tested are more selective for one or the other. If inhibition of PPGNs by both NE and 5-HT is important during sleep, these studies may provide a pharmacological basis for the choice of imipramine to treat nocturnal enuresis.

SUMMARY

Parasympathetic preganglionic discharges recorded from pelvic nerves in spinal cats were evoked by single-pulse stimulation of sacral afferent fibers or descending excitatory pathways in the thoracic spinal cord. Evoked responses were analyzed on-line by signal averaging. Discharges evoked by either pathway were increased in size by 2–10 times by coadministration of both picrotoxin and strychnine, but not by either drug alone. The combination also markedly enhanced evoked increases in bladder pressure. Strychnine was also effective with bicuculline and in cats depleted of central GABA stores by semicarbazide. The monoamine precursors, 5-HTP and L-dopa, only depressed evoked discharges in both untreated and convulsant-treated animals; this depression was enhanced by tricyclic antidepressants.

The results indicate that sacral parasympathetic preganglionic neurons are controlled by an unusual type of strong, local tonic inhibition which is mediated by both GABA and glycine. Both amino acid transmitters must be blocked to unmask this inhibition. Interruption of supraspinal control of this local inhibition may account in part for loss of bladder function following spinal injuries. The 5-HT and NE bulbospinal pathways that terminate near the preganglionic neurons also appear to be inhibitory. Enhancement of this inhibitory monoaminergic transmission by tricyclic antidepressants may contribute to the efficacy of these drugs in treating nocturnal enuresis.

ACKNOWLEDGMENTS

This work was supported in part by NIH grant HL-24085, and by a grant from The American Parkinson Disease Association.

REFERENCES

1. Barrington FJF: The relation of the hind-brain to micturition. Brain 44:23–53, 1921.
2. Bell JA, Anderson EG: The influence of semicarbazide-induced depletion of γ-aminobutyric acid on presynaptic inhibition. Brain Res 48:161–169, 1972.
3. Blackwell B, Currah J: The psychopharmacology of nocturnal enuresis. In Kolvin I, MacKeith R, Meadow SR (eds): "Bladder Control and Enuresis." London: Heinemann, 1973, pp 231–257.
4. Dahlström A, Fuxe K: Evidence for the existence of monoamine neurons in the central nervous system. II. Experimentally induced changes in the intraneuronal amine levels of bulbospinal neuron systems. Acta Physiol Scand (64 Suppl) 247:1–36, 1965.
5. Davidson N: "Neurotransmitter Amino Acids." London: Academic Press, 1976.
6. de Groat WC: The effects of glycine, GABA and strychnine on sacral parasympathetic preganglionic neurons. Brain Res 18:542–544, 1970.
7. de Groat WC: Nervous control of the urinary bladder of the cat. Brain Res 87:201–211, 1975.

8. de Groat WC: Mechanisms underlying recurrent inhibition in the sacral parasympathetic outflow to the urinary bladder. J Physiol (London) 257:503–513, 1976.
9. de Groat WC, Booth AM: Physiology of the urinary bladder and urethra. Ann Intern Med 92 (Part 2):312–315, 1980.
10. de Groat WC, Douglas JW, Krier J: The effects of L-5-hydroxytryptophan (5-HTP) on sacral parasympathetic reflex pathways to the large intestine and urinary bladder of the cat. Pharmacologist 21:266, 1979.
11. de Groat WC, Ryall RW: Reflexes to sacral parasympathetic neurons concerned with micturition in the cat. J Physiol (London) 200:87–108, 1969.
12. Franz DN, Neumayr RJ, Hare BD: L-dopa-induced hypotension: Depression of spinal sympathetic neurons by release of 5-hydroxytryptamine. Advan Exp Med Biol 90:219–232, 1977.
13. Kuru M: Nervous control of micturition. Physiol Rev 45:425–494, 1965.
14. Ryall RW, de Groat WC: The microelectrophoretic administration of noradrenaline, 5-hydroxytryptamine, acetylcholine and glycine to sacral parasympathetic preganglionic neurons. Brain Res 37:345–347, 1972.
15. Sangdee C, Franz DN: Enhancement of central norepinephrine and 5-hydroxytryptamine transmission by tricyclic antidepressants: A comparison. Psychopharmacology 62:9–16, 1979.
16. Shaffer D, Stephenson JD, Thomas DV: Some effects of imipramine on micturition and their relevance to its anti-enuretic activity. Neuropharmacology 18:33–37, 1979.

The Role of Peptides and Amino Acids as Neurotransmitters, pages 191–196
©1981 Alan R. Liss, Inc., 150 Fifth Avenue, New York, NY 10011

Effect of Systemically Administered Morphine on Spontaneous Activity of Globus Pallidus and Caudate Nucleus Neurons in the Rat

T. Celeste Napier and James H. Pirch

Both the globus pallidus (GP) and caudate nucleus (CN) have been shown to contain opiate receptors [Pert et al, 1976], and morphine is known to act at these receptors. In this study, we examined the effect of various doses of morphine, injected systematically, on the activity of spontaneously firing neurons in both the caudate nucleus and globus pallidus of the rat. A preparation was developed which allowed us to obtain a cumulative dose–response curve for each neuron.

MATERIALS AND METHODS

Adult Sprague–Dawley rats were anesthetized for surgery with sodium pentobarbital (50 mg/kg, ip) and supplemented with ether. Bilateral cannulas (16 ga for CN and 22 ga for GP) were permanently implanted on the dural surface directly above the area being studied. Stereotaxic coordinates were 0.8 mm anterior, 3 mm lateral for the GP and 2.6 mm anterior, 3 mm lateral for the CN, using bregma as zero [Pellegrino et al, 1979]. Earbar holders (PE tubing) were also permanently embedded into the implant, and the rats were allowed at least 1 week postsurgery recovery time. Sodium phenobarbital (140 mg/kg, ip) was chosen as the anesthetic for recording sessions, since this drug has a long duration of action and supplementations were rarely needed. Furthermore, a stable baseline firing rate could be maintained for hours after the initial injection (See Fig. 1).

Approximately 1 hr after phenobarbital administration, the animal was replaced into the stereotaxic instrument with the earbars inserted into the implanted earbar holders. This procedure circumvented the pain which would be produced

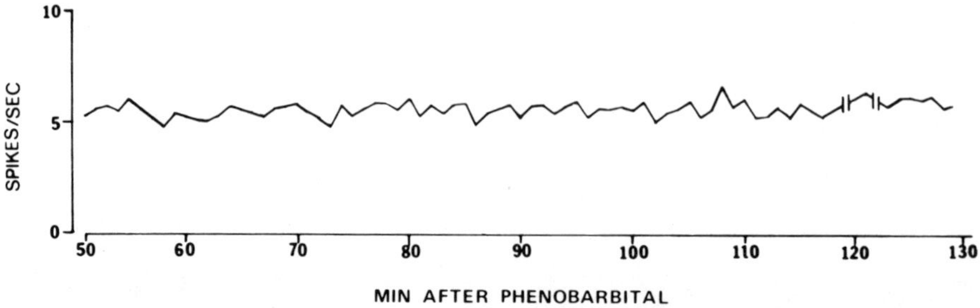

Fig. 1. Firing frequency of a neuron following a 100 mg/kg ip injection of sodium phenobarbital.

by insertion of earbars into the auditory canal. A metal microelectrode (either tungsten or stainless steel) was lowered through the guide cannula to the dorsal surface of either the CN or the GP. A spontaneously firing cell was then located, and activity was monitored throughout the extent of the experiment. The action potentials were displayed on a Tektronix 5111 storage oscilloscope after amplification via a Grass P15 microelectrode preamplifier. Individual spikes were isolated with a Frederick Haer amplitude analyzer, and the window output was fed into a Coulbourn R22-10 printing counter as well as a cumulative integrator. After a 15-min control period of stable activity, morphine sulfate was injected via a tail vein cannula in cumulative doses of 1, 5, 10, 20, and 30 mg/kg (as the salt) at 10-min intervals, and the neuron firing rate was determined for each 60-sec interval. Naloxone hydrochloride (1 mg/kg, iv) was administered at the end of each experiment. Naloxone antagonism, defined as the return of firing rate to control level, was used to verify morphine's action at opiate receptors. Control experiments were identical to the drug experiments, except saline was substituted (in equal volumes) for morphine. After each recording session, the site of recording was lesioned and electrode placement was verified histologically. Morphine dose–response curves were obtained for 24 neurons in the GP and 26 neurons in the CN. Five and six saline controls were performed in the GP and CN, respectively. Statistical significance of morphine's effect on single unit activity was determined by analysis of variance comparing treatment effects on the last five 1-min rates for each drug dose. A morphine effect was considered significant if the cell met two criteria: (1) at least a 15% change from control and (2) a significant F test ($P < 0.05$).

RESULTS

This animal preparation was found to be very suitable for electrophysiological recordings. The animals were only lightly anesthetized, and even though the higher doses of morphine did cause some respiratory depression, most of the animals maintained adequate respiratory function throughout the entire experimental period. The problem of synergisic interaction between morphine and the anesthetic on respiration was apparently reduced, since large doses of morphine were tolerated by the animals. The problem of drift in baseline firing rate due to the wearing off of short-acting anesthetics and the requirement for supplementation were also bypassed by the use of phenobarbital. The constant baseline rate is demonstrated by the saline controls (See Fig. 2). None of the saline controls met the criteria for a treatment effect, which strongly supports the view that the alteration of firing rate observed after treatment with morphine (and naloxone) was not due to an alteration of the baseline firing frequency during

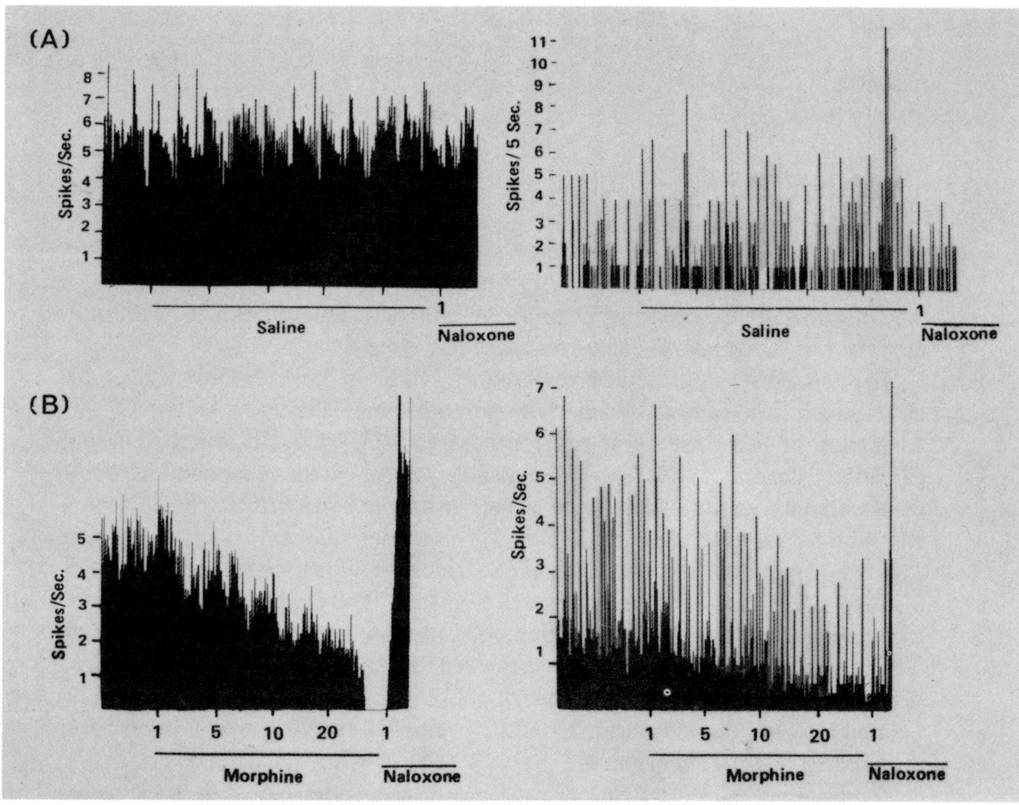

Fig. 2. Cumulative integrator output recordings of spontaneously firing single units. (A) Representative saline controls in GP (left) and CN (right). (B) Examples of GP (left) and CN (right) single units which demonstrated a significant depression of firing rate ($P < 0.05$) upon administration of cumulative doses of morphine.

TABLE I. Pattern of Neuronal Responses to Morphine

Response pattern	Total cells* Globus pallidus (%)	Caudate (%)
Depression Naloxone antagonized	75	27
Stimulation Naloxone antagonized	4	11
Depression Not antagonized	—	12
Stimulation Not antagonized	—	8
No change	21	42

*Globus pallidus, n = 24; caudate, n = 26.

the course of the experiment. Cumulative integrator recordings which illustrate the effect of morphine are also presented in Figure 2.

The general types of neuron response are given in Table I. Only three types of response to systemically injected morphine were observed in the GP: (1) depression of firing rate with naloxone antagonism, 75%; (2) stimulation with naloxone antagonism, 4%; and (3) no change, 21%. On the other hand, CN cells demonstrated a gamut of morphine effects on spontaneous firing rate: (1) depression with naloxone antagonism, 27%; (2) stimulation with naloxone antagonism, 11%; (3) depression by morphine but no naloxone antagonism, 12%; (4) stimulation by both morphine and naloxone, 8%; and (5) no change, 42%. Therefore, from the population of cells sampled, it would seem that more neurons in the GP responded to morphine with the expected depression as compared to the CN. A χ^2 analysis comparing the number of cells depressed and not depressed in the GP and CN was significant at $P < 0.005$, clearly verifying area differences in response to systemic morphine.

However, if the population of cells which showed a naloxone antagonized morphine depression is compared in the two areas, the dose–response relationships are very similar (Fig. 3). Both dose–response curves have significant linear regressions, indicative of a dose-related depression in firing rate. The ED_{50}s are similar, suggesting that for those cells which were depressed by morphine, the potency of morphine was the same in both areas.

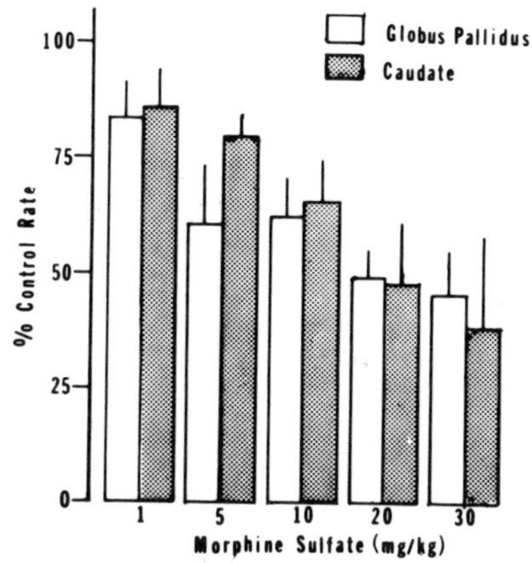

Fig. 3. Dose-related effects of morphine on globus pallidus and caudate neurons which demonstrated depression antagonized by naloxone. For globus pallidus, n = 18; linear regression $P < 0.01$; ED_{50} = 20 mg/kg. For caudate, n = 7; linear regression $P < 0.025$; ED_{50} = 23 mg/kg. ED_{50} was calculated from the linear regression and represents the dose which reduced firing rate to 50% control of.

DISCUSSION

In this paper the effects of systemically administered morphine on the firing of globus pallidus and caudate nucleus neurons were compared. In the dose range used in this experiment, morphine depressed a greater number of cells in the GP as compared to the CN (75 vs 27%, respectively). The effect of systemic morphine on spontaneously active CN cells has been studied in other laboratories [Kelmm and Mallari, 1978; Dafny and Rigor, 1980]. The authors are not aware of similar studies performed in the GP. However, since the major efferent projection of the CN is to the GP [Graybiel and Ragsdale, 1979; Nauta and Domesick, 1979], it is also appropriate to study the effects of systemically administered morphine in the GP. The depression elicted by morphine in the GP may be due not only to the action of morphine directly at the pallidal cell opiate receptor but also to indirect effects via morphine-induced modification of afferent input to the GP. This study does not allow determination of the specific site of action.

Attention should be drawn to the fact that when CN and GP neuron populations which show naloxone-antagonized morphine depression are compared, the sensitivities of the cells to morphine are identical (as indicated by the similar ED_{50}s). This finding could easily be missed if the investigator only compares overall response patterns and fails to compare similarly affected populations in the two separate nuclei.

The procedure described in this paper provides an experimental approach which appears to reduce some of the problems accompaning acute procedures. Stable firing rates are observed during the course of experiments of long duration. The doses of morphine tolerated in our animals are comparable and the CN neuron responses are similar to those in chronic experiments [Dafny and Rigor, 1980]. With this semichronic preparation one is able to locate cells with excellent signal-to-noise ratio so that the problem of background interference with the recorded single spike is usually eliminated. This procedure is presently being used in our laboratory to further characterize the mechanisms involved in morphine's effect on basal ganglia cells.

ACKNOWLEDGMENTS

The authors wish to thank Dr. Howard Strahlendorf for his consultation during the course of the study. This work was supported by USPHS BRSG RR05773 and MH29653 and by the Tarbox Parkinson's Disease Institute at Texas Tech University Health Science Center.

REFERENCES

1. Pert CB, Kuhar MJ, Snyder SH: Opiate receptor: Autoradiographic localization in rat brain. Proc Nat Acad Sci 73:3729–3733, 1976.
2. Pellegrino LJ, Pellegrino AS, Cushman AJ: "A Stereotaxic Atlas of the Rat Brain." New York: Plenum Press, 1979.
3. Klemm WR, Mallari CG: Morphine and naloxone effects on spontaneous unit activity in the caudate, central grey and amygdala. Prog Neuro-Psychopharmacol 2:535–542, 1978.
4. Dafny N, Rigor BM: Characterization of unit activity recorded from septum, thalamus and caudate following incremental opiate treatment. J Neurosci Res 5:177–127, 1980.
5. Graybiel AM, Ragsdale CW, Jr: Fiber connections of the basal ganglia. In Cuenod et al (eds): "Development and Chemical Specificity of Neurons." Prog Br Res 51:239–283, 1979.
6. Nauta JH, Domesick VB: The anatomy of the extrapyramidal system. In Fuxe K and Calne DB (eds): "Dopaminergic Ergot Derivatives and Motor Function." Oxford and New York: Pergamon Press, pp 3–22, 1979.

The Role of Peptides and Amino Acids as Neurotransmitters, pages 197–207
© 1981 Alan R. Liss, Inc., 150 Fifth Avenue, New York, NY 10011

Effects of Chronic Administration of Naltrexone on Appetitive Behaviors of Rats

I. M. Lang, J. C. Strahlendorf, H. K. Strahlendorf, and C. D. Barnes

Morphine-like stereospecific receptors in the brain were first reported by Goldstein et al in 1971. Hughes [1975] later extracted an endogenous peptide from mammalian brain which possessed morphine-like actions on guinea pig ileum and was inhibited by specific opioid antagonists. Recent evidence indicates that opiate receptor mechanisms may mediate appetitive behaviors. Systemic administration of specific opiate antagonists reduce deprivation-induced food [Holtzman, 1974; Brands et al, 1979; Brown and Holtzman, 1979; Frenk and Rogers, 1979; Holtzman, 1979; King et al 1979] and water [Holtzman, 1975; Maickel et al, 1977; Brown and Holtzman, 1979; Frenk and Rogers, 1979; Holtzman, 1979; Brown et al, 1980; Ostrowski et al, 1980] intake, hyperphagia of hypothalamic [King et al, 1979] or genetically [Margules et al, 1978] obese animals, and hyperdipsia caused by hypertonic saline [Brown et al, 1980; Brown and Holtzman, 1980; Czech and Stein, 1980] or angiotensin II [Summy-Long et al, 1980]. Also, intraventricular administration of morphine or β-endorphin increase food-deprived feeding in normal rats [Belluzzi and Stein, 1978; Kenney 1978]. The role of endogenous opiate mechanisms in noninduced or tonic feeding behavior is unclear because of a lack of experimental evidence [Brands et al, 1979; Tseng and Cheng, 1980] and cannot be inferred from prior experiments. In most of these studies, the animals were subjected to a variety of stressful stimuli including food or water deprivation, injection of drugs, surgery, and others. Recent evidence, however, suggest that stress-induced feeding is mediated through endogenous opiates [Lowy et al, 1980; McGivern and Berntson, 1980; Morley and Levine, 1980]. The observed inhibitory effects of opioid antagonists in prior studies, therefore, may have been due to a reduction of stress-related endogenous opioid release rather than the specific reduction of the appetitive behavior. Thus, we examined the role of endogenous opiates in tonic appetitive behavior using rats exposed to a minimum of stressful stimuli.

METHODS

Twelve 4-week-old Wistar–Kyoto rats anesthetized with ether were implanted subcutaneously with 2 cm long poly(ε-caprolactone) capsules and divided into two equal-sized groups. A control group of six rats received vehicle (ethyl oleate)-filled capsules, and the other group received naltrexone with vehicle. Filled capsules and drug release curve (Fig. 1) were obtained from Dr. Colin G. Pitt of the Research Triangle Institute, Research Triangle Park, North Carolina. Animals were housed individually in suspended wire-mesh cages in a temperature-controlled room (23–26°C) maintained on a 12-hr light–dark cycle (0700–1900). The animals were fed pelleted stock laboratory diet and tap water ad libitum, except during the deprivation periods. Food and water intake, urine output, and wet and dry fecal weight were measured daily. Body weight was measured weekly. Fluid output was calculated as the urine output plus the difference between the wet and dry fecal weights. Food and water intake were corrected to allow for the 10.5% water content of the stock diet. Twenty-four-hour food deprivation-induced feeding was tested over a period of 60 min, three times during the experiment. At the end of 8 weeks, the animals were subjected to an analgesic test in order to confirm the efficacy of the dose of naltrexone.

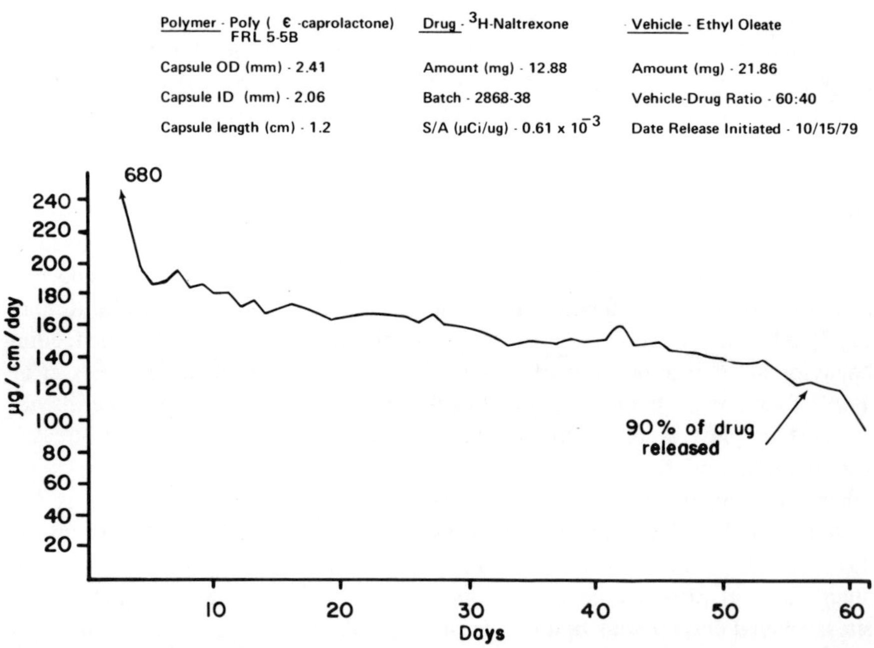

Fig. 1. Release curve of [^{3}H]naltrexone from poly(ε-caprolactone) capsule. Capsule and release curve obtained from Dr. Colin G. Pitt of the Research Triangle Institute, North Carolina.

The analgesic test was a hot-plate test (55 ± 2°C) using paw-lick as the end point.

Results are reported and displayed graphically as means ± SE. Lines were fitted to the means by regression analysis using BMDP programs (developed at the Health Sciences Computer Facility, UCLA, sponsored by NIH Special Resources Grant RR-3). Body weight was fitted to $\hat{y} = a + bx$ (BMD:P1R) and the other variables were fitted to $f = p_1 + p_2e^{p_3x}$ (BMD:P3R). The data obtained during the 24-hr deprivation periods and the subsequent 24 hr were eliminated from the regression analyses. Differences between control and experimental curves were compared by nonparametric analysis [Krauth, 1980]. When data points were compared, homogeneity of variances was confirmed using the F test before performing a two-tailed Students' t test [Snedecor and Cochran, 1972]. In all statistical tests, significance was tested at the 5% level.

RESULTS

The release curve of ^{3}H-naltrexone from a poly(ε-caprolactone) capsule immersed in physiologic saline is depicted in Figure 1. It is estimated that the 2-cm capsules released between 325 and 225 μg of naltrexone per day from Day 4 to Day 56 of the experiment.

Water intake (Fig. 2 and Table I) and fluid output (Fig. 3 and Table 1) of the

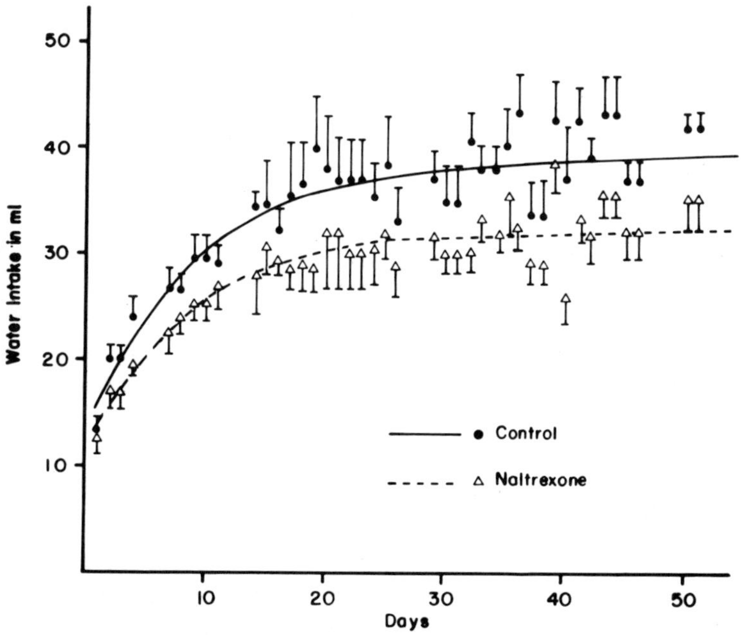

Fig. 2. Water intake of control and naltrexone-treated rats. Points are means and vertical bars are SE.

TABLE I. Regression Equations and Statistics of Metabolic Variables

Variable	n	Zero intercept (a or $p_1 - p_2$)	Sample regression coefficient (b)	Asymptote (p_1)	Curve shape (p_3)
Body weight[a]					
Control	6	46.1	4.39		
Naltrexone	6	42.7	4.18		
Food intake					
Control	6	6.23		19.8	−0.069
Naltrexone	6	6.66		19.3	−0.061
Dry fecal weight					
Control	6	1.17		4.15	−0.088
Naltrexone	6	0.99		4.01	−0.084
Water intake					
Control	6	6.24		38.88	−0.111
Naltrexone	6	6.25		32.34*	−0.115
Fluid output					
Control	6	−0.17		18.72	−0.082
Naltrexone	6	0.67		15.63*	−0.075

[a]Body weight fitted to $\hat{y} = a + bx$ and other variables fitted to $f = p_1 + p_2e^{p_3x}$; n, number of animals.
*$P < 0.05$ for differences between control and naltrexone groups.

naltrexone-treated group were significantly reduced compared to the control group. The asymptotes of the curves were reduced to similar extents, 17%. No differences were observed in the zero intercepts or curve shapes.

Naltrexone treatment had no effect on food intake (Fig. 4 and Table I), dry fecal weight (Fig. 5 and Table I), body weight (Fig. 6, Table I), or food deprivation-induced feeding (Table II). Paw-lick latency was significantly reduced in the naltrexone-treated group (Table III).

DISCUSSION

The reduction in water consumption caused by chronic administration of the opiate antagonist, naltrexone, in these minimally stressed rats suggests that endogenous opiates may mediate tonic fluid ingestion. In the two studies reported to date that investigated the effects of chronic administration of opioid antago-

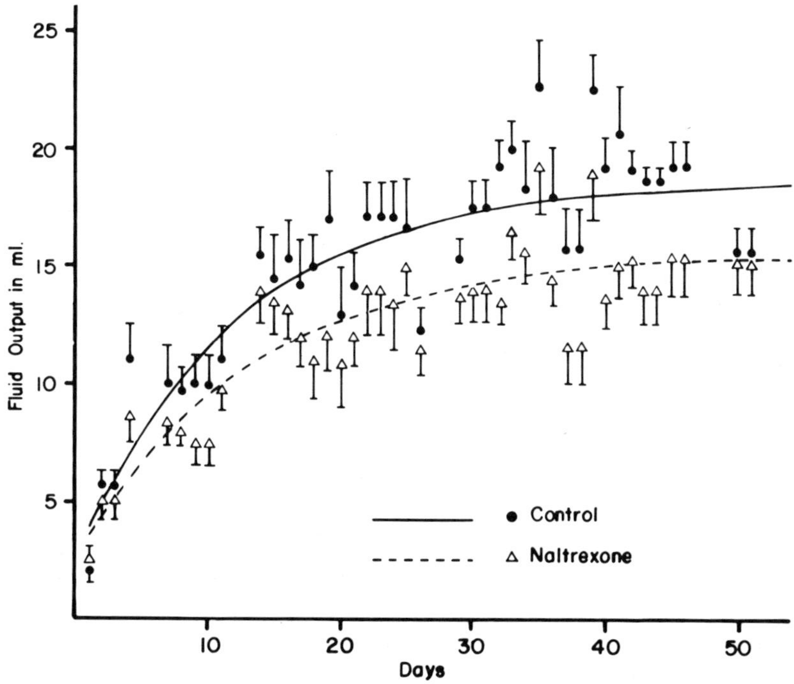

Fig. 3. Fluid output of control and naltrexone-treated rats. Points are means and vertical bars are SE.

TABLE II. Twenty-four Hour Food Deprivation-Induced Feeding Test*

Weeks after implantation	Grams of food per 100 g of body weight		P
	Control	Naltrexone	
2	1.2 ± 0.1 (6)	1.2 ± 0.1 (6)	NS
4	1.7 ± 0.1 (6)	1.7 ± 0.2 (6)	NS
7	1.7 ± 0.1 (6)	1.9 ± 0.2 (6)	NS

*Amount eaten in 60 min. Values are means ± SE based on one observation in each of (n) animals. P, probability that the means of control and naltrexone groups differ; NS, not significantly different.

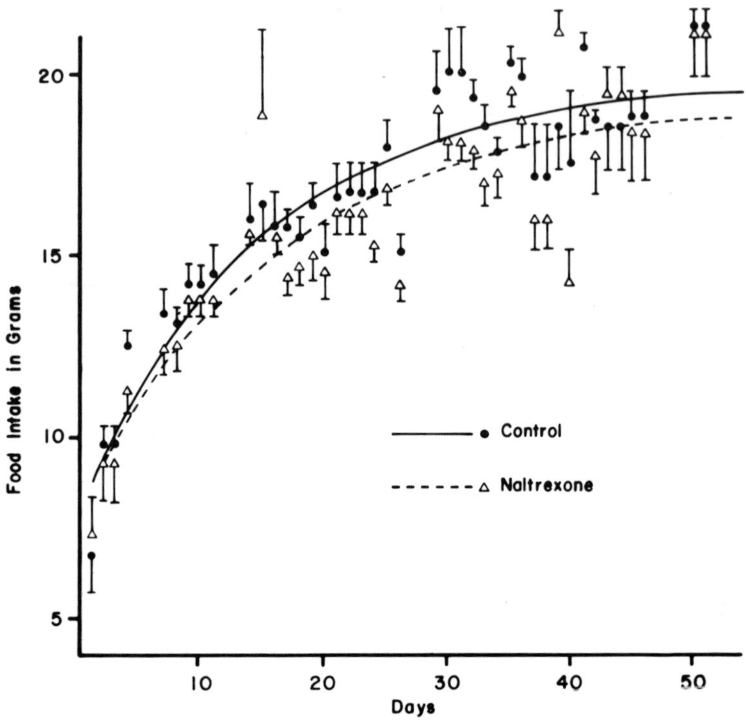

Fig. 4. Food intake of control and naltrexone-treated rats. Points are means and vertical bars are SE.

nists, reduction of food and water intake had been found through only the first or second day after antagonist administration [Brands et al, 1979; Tseng and Cheng, 1980]. Since administration of the antagonist probably stressed the animals (ie, subcutaneous injection of 2 ml of zinc tannate salt of naloxone (18 mg/kg), or subcutaneous implantation of four 2-mg pellets of naltrexone), the observed short-lasting appetite suppression may reflect a reduction of stress-related ingestion rather than the reduction of tonic appetitive behavior. The narcotic antagonist, naloxone, reduced stress-related feeding induced by tail-pinch, glucoprivic agents, or deprivation [Lowy et al, 1980; Morley and Levine, 1980]. The rate of administration of the antagonist in both experiments was unknown. Therefore, another possible explanation for the short-lasting appetite suppression is that naloxone was released rapidly and was dissipated within 2 days. Our rats were subjected to surgical stress initially but received no subsequent injections or painful stimuli. The three 24-hr deprivation periods were separated by at least 2 weeks. Under these conditions, it is unlikely that the

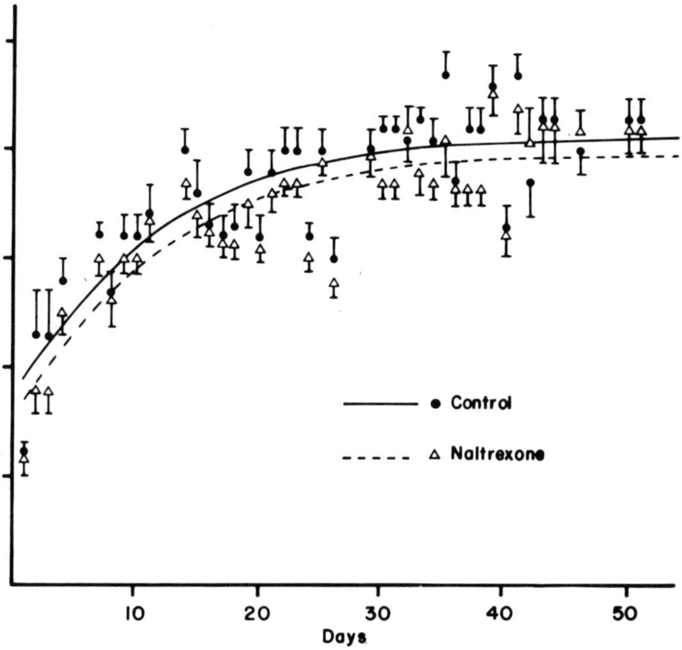

Fig. 5. Dry fecal weight of control and naltrexone-treated rats. Points are means and vertical bars are SE.

TABLE III. Analgesic Test

Group	Paw-lick latency[a]	t
Control	8.2 ± 0.7 (5)	4.49*
Naltrexone	5.0 ± 0.3 (6)	

[a]Values are means ± SE based on four observations in each of (n) animals. t, Students' t statistic.
*$P < 0.01$.

suppression of fluid intake we observed was related to reduction of stress-induced ingestion.

The finding that water intake and fluid output were reduced to the same extent (17%) suggests that osmoregulatory mechanisms were not preferentially affected by naltrexone. Osmoregulatory mechanisms, however, can be affected by opiate antagonists. Hypertonic saline-induced drinking was reduced by prior administration of naloxone [Brown and Holtzman, 1980; Brown et al, 1980; Czech and Stein, 1980]. This reduction of osmoregulatory drinking may be part of a general

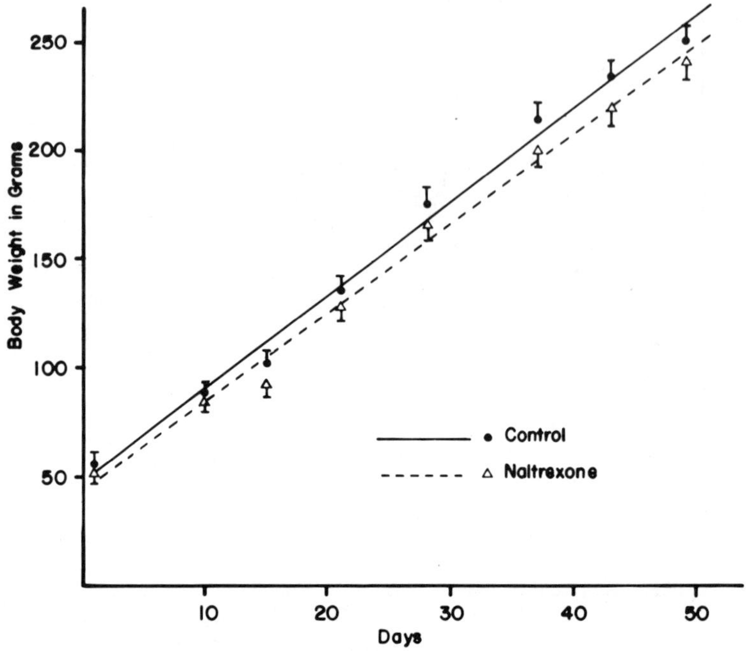

Fig. 6. Body weight of control and naltrexone-treated rats. Points are means and vertical bars are SE.

reduction of all appetitive behaviors. The administration of opioid antagonists suppressed all appetitive behaviors thus far examined [Holtzman, 1974; Maickel et al, 1977; Margules et al, 1978; Brands et al, 1979; King et al, 1979; Brown et al, 1980; Lowy et al, 1980; Summy-Long et al, 1980; Tseng and Cheng, 1980].

The suppressant effect of opioid antagonists on appetitive behavior has been attributed to the role of endogenous opiates in mediation of drive-reduction reward [Belluzzi and Stein, 1977]. That is, opiate antagonists reduce ingestive behavior by inhibiting the endogenous opiate-mediated reward associated with food and water intake. This may explain the ability of opioid antagonists to suppress all types of appetitive behaviors, but it does not account for all the effects on ingestion. Opioid antagonists suppress drinking more than feeding. The percentage of reduction in intake is greater and the threshold dose is lower for deprivation-induced drinking than feeding [Brown and Holtzman, 1979; Frenk and Rogers, 1979; Holtzman, 1979]. We also found a differential sensitivity of tonic ingestive behavior to opioid antagonist. Tonic drinking but not tonic (or deprivation-induced) feeding was significantly reduced by the dose of naltrexone used in our experiments. The drive-reduction reward theory of Belluzzi and Stein [1977] does not explain this differential sensitivity of ingestive behavior. Perhaps in addition to drive-reduction reward, the endogenous opiates mediate a specific mechanism in the control of fluid homeostasis. In this regard, it is interesting to note that Brattleboro rats (genetically deficient in antidiuretic hormone, ADH) are more resistant to pain and do not exhibit cold-water swim analgesia [Bodnar et al, 1980]. Both phenomena are reversed by systemic administration of arginine-vasopressin [Bodnar et al, 1980], suggesting that ADH mediates analgesic mechanisms. Hypophysectomy does not alter cold-water swim analgesia [Glusman et al, 1979] or opiate antagonist suppression of deprivation- or hypertonic saline-induced drinking [Brown et al, 1980]. Therefore if ADH is related to analgesia and drinking, its actions are not mediated through the pituitary gland but probably through the hypothalamus. Antidiuretic hormone is synthesized in hypothalamic magnocellular neuroendocrine cells [Hayward, 1971]. Since the hypothalamus contains centers mediating drinking behavior [Andersson, 1978; Johnson and Buggy, 1978], and has high levels of endorphin [Krieger and Liotta, 1979] and ADH [Hayward, 1977; Swanson, 1977], it does not seem unreasonable to suggest that opiate antagonists may preferentially affect drinking behavior by blocking interactions among these hypothalamic centers. One such interaction has been examined recently. The suppression of angiotensin II-induced release of ADH caused by cerebroventricular administration of Leu-enkephalin was reversed by prior subcutaneous administration of naltrexone [Summy-Long et al, 1980].

SUMMARY

Opioid antagonists have been reported to reduce water and food intake. In the experiments, ingestive behavior was sometimes induced using stressful stimuli. These experiments were conducted to determine the effects of chronic administration of an opioid antagonist on appetitive behaviors of minimally stressed rats. Twelve 4-week old rats were randomly separated into two equal-sized groups, anesthetized with ether, and implanted subcutaneously with poly(ε-caprolactone) capsules. One group received naltrexone in the capsules, the other group received vehicle. These capsules released the drug at a rate of approximately 250 μg/day for over 8 weeks. Gross metabolic behavior (food intake, water intake, urine output, fecal output, and body weight) was monitored for 8 weeks. At three times during the 8 weeks, food deprivation (24 hr)-induced feeding was tested over a period of 60 min. At the end of the eighth week, the animals were subjected to an analgesic test in order to confirm the efficacy of the naltrexone dose. No differences in food intake, body weight, fecal output, or deprivation-induced feeding were observed between the control and naltrexone groups. Water intake and fluid output of the naltrexone group were significantly reduced compared to the control group. These results suggest that endogenous opioids may mediate mechanisms of tonic drinking behavior and cannot be explained as representing effects on stress-induced ingestion.

ACKNOWLEDGMENTS

This work was supported in part by HLO7289 and Tarbox Parkinson Disease Institute.

REFERENCES

1. Andersson B: Regulation of water intake. Physiol Rev 58:582–603, 1978.
2. Belluzzi JD, Stein L: Enkephalin may mediate euphoria and drive-reduction reward. Nature (London) 266:556–558, 1977.
3. Bodnar RJ, Kelly DD, Brutus M, Glusman M: Stress-induced analgesia: Neural and hormonal determinants. Neurosci Biobehav Rev 4:87–100, 1980.
4. Brands B, Thornhill JA, Hirst M, Gowdey CW: Suppression of food intake and body weight gain by naloxone in rats. Life Sci 24:1773–1778, 1979.
5. Brown DR, Blank MS, Holtzman SG: Suppression by naloxone of water intake induced by deprivation and hypertonic saline in intact and hypophysectomized rats. Life Sci 26:1535–1542, 1980.
6. Brown DR, Holtzman SG: Suppression of deprivation-induced food and water intake in rats and mice by naloxone. Pharmacol Biochem Behav 11:567–573, 1979.
7. Brown DR, Holtzman SG: Evidence that opiate receptors mediate suppression of hypertonic saline-induced drinking in the mouse by narcotic antagonists. Life Sci 26:1543–1550, 1980.
8. Czech DA, Stein EA: Naloxone depresses osmoregulatory drinking in rats. Pharmacol Biochem Behav 12:987–989, 1980.
9. Frenk H, Rogers GH: The suppressant effects of naloxone on food and water intake in the rat. Behav Neural Biol 26:23–40, 1979.
10. Goldstein A, Lowney LI, Pal BK: Stereospecific and non-specific interactions of the morphine narcotic congener levorphanol in subcellular fractions of mouse brain. Proc Natl Acad Sci 68:1742–1747, 1971.
11. Glusman M, Bodnar RJ, Kelly DD, Sirio C, Stern J, Zimmerman EA: Attenuation of stress-induced analgesia by anterior hypophysectomy in the rat. Neurosci Abst 5:609, 1979.
12. Hayward JN: Functional and morphological aspects of hypothalamic neurons. Physiol Rev 57:574–658, 1977.
13. Holtzman SG: Behavioral effects of separate and combined administration of naloxone and d-amphetamine. J Pharmacol Exp Ther 189:51–60, 1974.
14. Holtzman SG: Effects of narcotic antagonists on fluid intake in the rat. Life Sci 16:1465–1470, 1975.
15. Holtzman SG: Suppression of appetitive behavior in the rat by naloxone: Lack of effect of prior morphine dependence. Life Sci 24:219–226, 1979.
16. Hughes J: Isolation of an endogenous compound from the brain with pharmacological properties similar to morphine. Brain Res 88:295–308, 1975.
17. Johnson AK, Buggy J: Periventricular preoptic-hypothalamus is vital for thirst and normal water economy. Am J Physiol 234:R122–R129, 1978.
18. Kenney NJ, McKay LD, Woods SC: Effect of intraventricular β-endorphin on food intake in rats. Neurosci Abstr 4:176, 1978.
19. King BM, Castellanos FX, Kastin AJ, Berzas MC, Mauk MD, Olson GA, Olson RD: Naloxone-induced suppression of food intake in normal and hypothalamic obese rats. Pharmacol Biochem Behav 11:729–732, 1979.

20. Krieger DT, Liotta AS: Pituitary hormones in brain: Where, how and why? Science 205:366–372, 1979.
21. Lowy MT, Maickel RP, Yim GKW: Naloxone reduction of stress-related feeding. Life Sci 26:2113–2118, 1980.
22. McGivern RF, Berntson GG: Mediation of diurnal fluctuations in pain sensitivity in the rat by food intake patterns: Reversal of naloxone. Science 210:210–211, 1980.
23. Maickel RP, Braude MC, Zabik JE: The effects of various narcotic agonists and antagonists on deprivation-induced fluid consumption. Neuropharmacology 16:863–866, 1977.
24. Margules DL, Moisset B, Lewis MJ, Shibuya H, Pert CB: Beta-endorphin is associated with overeating in genetically obese mice (ob/ob) and rats (fa/fa). Science 202:988–991, 1978.
25. Morley JE, Levine AS: Stress-induced eating is mediated through endogenous opiates. Science 209:1259–1261, 1980.
26. Ostrowski NK, Foley TI, Lind MD, Reid LD: Naloxone reduces fluid intake: Effects of water and food deprivation. Pharmacol Biochem Behav 12:431–435, 1980.
27. Snedecor GW, Cochran WG: "Statistical Methods." Ames, Iowa: State University Press, 1972.
28. Summy-Long J, Keil LC, Deen K, Severs W: Opiate inhibition of angiotensin drinking and vasopressin release. Fed Proc 39:762, 1980.
29. Swanson LW: Immunohistochemical evidence on a neurophysin containing autonomic pathway arising in the paraventricular nucleus of the hypothalamus. Brain Res 128:346–353, 1977.
30. Tseng LF, Cheng DS: Acute and chronic administration of β-endorphin and naloxone on food and water intake in rats. Fed Proc 39:606, 1980.

The Role of Peptides and Amino Acids as Neurotransmitters, pages 209–215
© 1981 Alan R. Liss, Inc., 150 Fifth Avenue, New York, NY 10011

The Use and Abuse of Clonidine as a Psychopharmacological Tool

Abraham Flemenbaum

Clonidine (CLO), an imidazoline derivative, was synthesized by Stahle in 1962 and is utilized clinically mainly as an antihypertensive.

Early in the evaluation of the mechanism of action of CLO in reducing blood pressure, Kobinger [1973] already concluded that the sympathomimetic effects of CLO are due to a direct action of the drug upon adrenergic sites by stimulation of α-receptors. Clonidine's actions are similar to those reported for other imidazolines such as naphazoline and tetrahydrazoline in that most of its actions are effectively blocked by α-adrenergic receptor inhibitors and in that CLO does not exert β-adrenergic stimulating properties. Kobinger [1973] also reported that no stimulation of central 5-hydroxytryptamine and/ or dopamine receptors had been found in previous studies.

Because of its specific α-adrenergic properties and its ability to cross the blood–brain barrier, CLO has been extensively used as a psychopharmacological tool for the evaluation of postsynaptic of norepinephrine (NE) agonist effects. Psychoneuroendocrinology studies have been reported in disparate areas; CLO has been shown to release growth hormone (GH) [Lal et al, 1975] and has been used to measure the increased sensitivity induced by electro-convulsive shock at the postsynaptic catecholamine receptors in animals; ie, the psychomotor "stimulant" effects of CLO and/or apomorphine given after reserpine were enhanced by pretreatment with electro-convulsive treatment [Modigh, 1975]. Matussek et al [1977] used the drug as a measure of α-adrenergic receptor sensitivity in depressed patients and controls. At one time it gained such a degree of respectability that even well-recognized researchers such as Mendels and Frazier used it as one more research drug in psychoneuroendocrinology for its adrenergic agonistic properties [Mendels et al, 1977].

For a long time, CLO was known as a "specific NE receptor site agonist," [Kobinger, 1973; Schmitt and Schmitt, 1969; Andén et al, 1970; Corrodi et al,

1970]. It was also known—but disregarded—that its behavioral effects, ie, the hyperactivity expected from central NE activation, is not manifested unless it is combined with other drugs such as reserpine [Andén and Strömbom, 1974] and that CLO itself does not stimulate locomotor activity [Fuxe and Understedt, 1970; Maj et al, 1972]. Also contrary to its specificity for a NE receptor site, other studies reported that CLO can hardly be regarded as a specific central NE agonist [Maj et al, 1975].

Instead of a synergism with antidepressants, all the tricyclics have been reported to reduce the antihypertensive effects of CLO [Van Spanning and Van Zwieten, 1973]. Also, even though CLO was initially reported as not producing depression in humans as some other antihypertensives do, a recent report [Roberts, 1977] indicates that "CLO is well recognized now as a cause of depression when used to treat hypertension," confirming the author's own observations of unfavorable interactions between CLO and tricyclics in two cases. Related to the (heuristic) theory of dopamine receptors in schizophrenia, a recent report that pimozide—formerly thought to be a neuroleptic specific for dopamine blockade—is a potent inhibitor of a NE-stimulated adenylate cyclase, prompted the speculation that NE blockade may be relevant to antipsychotic effects [Blumberg et al, 1975]. This is in spite of the fact that α-NE blockers like phenoxybenzamine and the beta-blocker propranolol have little or no antipsychotic properties and CLO, assumed to be an α-noradrenergic agent, does not produce psychosis [Melzer and Stahl, 1976].

Furthermore, in our laboratory we studied eight groups of five animals each that received CLO in a chronic pattern within a procedure to evaluate the interaction of α- and β-adrenergic drugs and lithium. The results were as discouraging as they were unreliable. Some animals presented biphasic effects of CLO with a slight initial hyperactivity and then sedation or the reverse. There were no consistent results on duplication of the experiment (see Table I), and our observations of the CLO effects in the animals (ie, decrease of locomotor activity-sedation simultaneously with peripheral sympathomimetic symptoms) were astonishingly similar to the ones reported by Maj and collaborators [1975]. These results and the reports mentioned previously cast doubt about the usefulness of CLO as a postsynaptic NE research tool. However, for a while at least, reports continued to proliferate based on the assumption that CLO is a specific α-NE agonist.

More recent reports, however, more clearly point in other directions, although CLO does appear to have some psychotropic effects. Björkqvist [1975] in a double-blind study described that CLO was useful in the treatment of alcohol withdrawal, especially "when it was desirable to minimize the use of tranquilizers" as it decreased the tremors, sweating, the elevated systolic blood pressure, the anxiety, and depression. CLO has also been reported to inhibit naloxone-

TABLE I. Rat Activity[§] by Experimental Condition[†] on Initial Experiment (1) and Duplication of the Same (2) (in a fresh group of animals)

	Condition							
	1 (Saline)	2 (Clonidine)[e]	3 (Rest)	4 (Test Drug)	5 (Rest)	6 (Combination of 2 drugs)	4 vs 2	6 vs 2
(1)	95 ± 69	189 ± 132	100 ± 50	(a_1) 39 ± 22*[a]	(a_1)	318 ± 214*	*	**
(2)	263 ± 27	243 ± 126	500 ± 161**	(a_2)196 ± 40**[a]	(a_2)	498 ± 211*	NS	**
(1)	111 ± 57	357 ± 95***	160 ± 120	(b_1) 48 ± 36**[b]	241 ± 146	182 ± 128	***	**
(2)	144 ± 88	95 ± 85	252 ± 153**	(b_2)150 ± 107[b]	212 ± 139**	106 ± 77	NS	NS
(1)	101 ± 65	279 ± 116**	341 ± 205**	(c_1)211 ± 115[c]	221 ± 262	319 ± 123**	NS	***
(2)	139.2 ± 36	210 ± 108	245 ± 59**	(c_2)164 ± 40[c]	144 ± 65	267 ± 176	NS	NS
(1)	114 ± 27	216 ± 119*	177 ± 45*	(d_1)159 ± 77[d]	163 ± 101	352 ± 128**	NS	**
(2)	247 ± 133	242 ± 137	433 ± 320*	(d_2)389 ± 341[d]	454 ± 400	375 ± 132	NS	NS

§For stereotyped behavior the same unduplicable, unreliable results were observed.

†Procedure: Male Sprague–Dawley rats of 150 mg initial weight in a normal daylight photoperiod were placed in activity wheels each day from 14:00 to 16:00 hr and were given 1 week of acclimation to the lab, 1 week of acclimation to activity wheels, 1 week of handling, 1 week of saline (col. 1), (2) 2 weeks of clonidine 0.4 mg/kg, (3) 1 week of rest, (4) 2 weeks of the test drug, (5) 1 week of rest (or continuation of lithium when lithium was the test drug), and (6) 1 week of combinations of both drugs. Activity was measured as number of wheel turns ± standard deviation. Statistical evaluation was done with t tests utilizing pooled variances comparing each condition versus the own saline control unless specified otherwise (ie, condition 4 vs 2). Test drugs utilized were given daily ip 10 min before measurements (except for lithium which was given twice daily at 10:00 and 16:30 hr) as follows:

[a]Lithium 1.5 mEq/kg.

[b]Dybenzyline 2.0 mg/kg.

[c]Propranolol 5.0 mg/kg.

[d]Methisergide 3.0 mg/kg.

[e]Animals on clonidine were consistently observed to be abnormally sedated with occasional bursts of hyperactivity, piloerection, and abnormal urinary and fecal excretion.

*P < 0.1; **P < 0.05; ***P < 0.01.

(1) = First experiment; (2) = Duplication.

induced withdrawal signs in morphine-dependent rats [Tseng et al, 1975] and in humans [Washton et al, 1980].

As mentioned above, CLO may produce depression when used as an antihypertensive medication, and it appears to have antimanic effects [Cohen 1980], as well as appearing useful in Tardive Dyskinesia and Guiles de la Tourette syndrome [Jouvent et al, 1980; Cohen et al, 1979]. Thus, it is understandable that CLO has become an important addition to the armamentarium of drugs utilized for psychopharmacology research.

However, there are still certain findings that are of importance and that have resulted in a reevaluation regarding the validity of studies utilizing CLO as a "pure" postsynaptic α-adrenergic agonist substance. Recently, Dollery [1976] again concluded that experimental work in animals has shown that CLO lowers blood pressure prominently by an action in the CNS due to α-adrenoceptor stimulation. Dollery [1976] called attention to previous studies demonstrating that it may also diminish the release of noradrenaline from peripheral nerves by stimulating prejunctional α-receptors concerned with modulating the amount of transmitter released from the neuron as a result of the nerve impulse. Furthermore, although CLO reduces plasma and urinary catecholamines, he stated that it appears to have a biphasic effect depending on the dose, as two patients who reportedly took an overdose of 4–5 mg of CLO both had prolonged hypertension.

Although his observation theoretically could be explained by an increased stimulation of the peripheral α-receptors, as early as 1970 and 1976 a comprehensive review [Scriabine et al, 1976] gave a different insight into the mechanism of action of CLO. In addition to the previously described effects of CLO on the central nervous system, Scriabine [1976] reported inhibition of sympathetic transmission at peripheral sites, ie, reduction of the heart rate in dogs with a spinal cord section at the level of the cervical vertebrae. In cats pretreated with reserpine or AMPT at doses sufficient to produce the usual CLO inhibitory effect on sympathetic nerve activity. In isolated rabbit hearts, CLO also antagonized the release of NE caused by electrical stimulation of intact sympathetic nerves. Furthermore, the destruction of noradrenergic neurons by 6-hydroxydopamine reduced the antihypertensive effects of CLO in rabbits. CLO was found to diminish the stimulation-evoked overflow from slices of rats' cerebral cortex preincubated with tritiated NE. The α-adrenergic blocking agents, phenotolamine and phenoxybenzamine, increased the stimulation-induced release of NE from cerebral cortex slices [Starke 1973; Scriabine et al, 1976; Buckman, 1976].

These earlier researchers hypothesized, and more recent reports clearly conclude [Isaac 1980], that the presynaptic α-adrenergic receptor is specifically stimulated as one of CLOs pharmacological actions. Thus, the release of NE by nerve impulses is inhibited by this feedback system; the feedback loop consists of released NE and α-adrenergic receptors at the presynaptic nerve endings, with pharamcological effects clearly the reverse of the ones hypothesized earlier.

Also, α-adrenergic antagonists block the access of NE and other α-stimulants (ie, CLO) to the receptors at the nerve membrane, thereby interrupting the presynaptic feedback loop and facilitating the release of NE.

This observation raises the question of the importance of the prejunctional α-adrenergic receptors to the central antihypertensive effect of CLO [Isaac, 1980]. Therefore, the evidence is strong that the actions of CLO may not all be a direct stimulation of postsynaptic α-adrenergic receptors but also involve a biphasic stimulation of presynaptic receptors interfering with the NE feedback system and turnover, partial agonism and even antagonism of postsynaptic α-receptors [Isaac, 1980].

This concept makes more understandable the CLO antidepressant interaction. If CLO acted simply by potentiating NE at the postsynaptic level, a mutual potentiation or synergism would be expected with tricyclic antidepressants and no depression should be reported from the use of CLO. On the contrary, there is an antagonism of the hypotensive effects of CLO by tricyclics and as mentioned above depression has also been reported in patients maintained on it. This interaction of CLO with tricyclics as well as CLOs possible antimanic effects is of clinical and theoretical importance because it further supports a presynaptic site of action for CLO.

In brief, the author has reviewed the literature that is indicative that the popularity of CLO as a research tool in psychopharmacology as a central postsynaptic adrenoceptor agonist is based on a simplistic and incomplete understanding of its mechanism of action and that a complete reevaluation is needed regarding its prejunctional and agonistic properties and also even its controversial partial antagonism at postsynaptic receptors.

SUMMARY

Because early reports on the mechanisms of action of clonidine emphasized its "pure" postsynaptic α-norepinephric stimulatory properties, clonidine has enjoyed an ever-growing popularity in psychoneuroendocrinology research as a central NE agonist.

Clonidine appears to have some psychotropic activity. However, its biphasic and sometimes unpredictable effects (such as the ones we obtained in our laboratory) and the evidence of activity at presynaptic receptors which interfers with the NE feedback loop and turnover indicates that the rationale for its use in research is in need of complete reevaluation and may be even opposite to the one heretofore utilized.

ACKNOWLEDGMENTS

This paper was made possible in part by research grants from the Southern Medical Association and the Medical Foundation of Texas Tech University Health Sciences Center. Clonidine was gracefully supplied by Boehringer Ingelheim Ltd.; dibenzyline by Smith, Kline, and French Labs, Philadelphia, Pennsylvania; propranolol by Ayerst Laboratories Incorporated, New York, New York; and Methysergide by Sandoz, Incorporated, East Hanover, New Jersey.

REFERENCES

1. Andén NE, Corrodi H, Fuxe K, Hökfelt B, Hökfelt T, Rydin C, Svensson T: Evidence for a central noradrenaline receptor stimulation by clonidine. Life Sci 9:513–523, 1970.
2. Andén NE, Strömbom U: Adrenergic receptor blocking agents: Effects on central noradrenaline and dopamine receptors and on motor activity. Psychopharmacologia (Berlin) 38:91–103, 1974.
3. Björkqvist SE: Clonidine in alcohol withdrawal. Acta Psychiat Scand 52:256–263, 1975.
4. Blumberg JB, Taylor RE, Sulser F: Blockade by pimozide of a noradrenaline sensitive adenylate cyclase in the limbic forebrain: Possible role of limbic noradrenergic mechanisms in the mode of action of antipsychotics. J Pharm Pharmacol 27:125–128, 1975.
5. Buckman RW: Drug interactions and the practice of psychiatry. Curr Concepts Psychiat 2:1, 1976.
6. Cohen DJ, Young JG, Nathanson JA, Shaywitz BA: Clonidine in Tourette's Syndrome. Lancet 2:551–553, 1979.
7. Cohen Sidney: Clonidine: Non-opiate detoxification—Drug Abuse and Alcoholism. Newsletter IX: 6, 1980.
8. Corrodi H, Fuxe K, Ljungdahl A, Ögren S: Studies on the action of some psychoactive drugs on central noradrenaline neurones after inhibition of dopamine-b-hydroxylase. Brain Res 24:451–470, 1970.
9. Dollery CT: Centrally acting alpha-adrenoceptor agonists in hypertension: Mechanisms and their role in therapy. Aust NZJ Med (Suppl. 3) 6:88–94, 1976.
10. Fuxe K, Ungerstedt U: Histochemical, biochemical and functional studies on central monoamine neurons after acute and chronic amphetamine administration. In Costa E and Garattini S (eds): "Amphetamines and Related Compounds." New York: Raven Press, 1970, pp 257–288.
11. Isaac L: Clonidine in the central nervous system: Sites and mechanism of hypotensive action. J Cardiovasc Pharmacol (Suppl. 1) 2:S5–19, 1980.
12. Jouvent R, Lecrubier Y, Puech AJ, Simon P, Widlöcher D: Antimanic effect of clonidine. Am J Psychiat 137:1275–1276, 1980.
13. Kobinger W: Pharmacologic basis of the cardiovascular actions of clonidine. In Onesti et al (eds): "Hypertension: Mechanisms and Management." New York: Grune & Stratton, 1973, pp 369–380.
14. Lal S, Tolis G, Martin JB, Brown GM, Guyda H: Effect of clonidine on growth hormone, prolactin, luteinizing hormone, follicle-stimulating hormone and thyroid-stimulation hormone in the serum of normal men. J Clin Endocrinol Metabol 41:827–832, 1975.
15. Maj J, Sowinska H, Baran L, Kapturkiewicz Z: The effect of clonidine on locomotor activity in mice. Life Sci 11:483–491, 1972.
16. Maj J, Baran L, Sowinska H, Zielinski M: Influence of cholinolytics on clonidine action. Pol J Pharmacol Pharm 27:17–26, 1975.
17. Matussek N, Ackenheil M, Hippius H, Schroder H-T, Schultes H, Wasilewski B: Effect of clonidine on HGH release in psychiatric patients and controls. Proceedings of VI World Congress of Psychiatry, Honolulu, Hawaii. p 96.
18. Meltzer HY, Stahl SM: The dopamine hypothesis of schizophrenia: A review. Schizophrenia Bul 2:19–76, 1976.
19. Mendels J, Maany I, Brunswick D, Frazer A, Ramsey TA: Growth hormone responses in depression. Proceedings of VI World Congress of Psychiatry, Honolulu, Hawaii. p. 96.

20. Modigh K: Electroconvulsive shock and postsynaptic catecholamine effects: Increased psychomotor stimulant action of apomorphine and clonidine in reserpine pretreated mice by repeated ECS. J Neurol Transmission 36:19–32, 1975.
21. Roberts JA: Dixarit (clonidine hydrochloride) and depression. Med J Austral 1:158, 1977.
22. Schmitt H, Schmitt MH: Localization of the hypotensive effect of 2-(6-dichlorophenylamino)-2-imidazoline hydrochloride (ST 155, catapresan). Eur J Pharmacol 6:8–12, 1969.
23. Scriabine A, Clineschmidt BV, Sweet CS: Central noradrenergic control of blood pressure. Ann Rev Pharmacol Toxicol 16:113–123, 1976.
24. Starke K: Regulation of catecholamine release: α-receptor mediated feed-back control in peripheral and central neurones. In Usdin E and Snyder SH (eds): "Frontiers in Catecholamine Research." New York: Pergamon Press, 1973, pp 561–565.
25. Tseng L-F, Loh HH, Wei ET: Effects of clonidine on morphine withdrawal signs in the rat. Eur J Pharmacol 30:93–99, 1975.
26. Van Spanning HW, Van Zwieten PA: Interference of tricyclic antidepressants with the central hypotensive effect of clonidine. Eur J Pharmacol 24:402–404, 1973.
27. Washton AM, Resnick RB: Clonidine for opiate detoxification—outpatient clinical-trials. Am J Psychiat 137:1121–1122, 1980.

The Role of Peptides and Amino Acids as Neurotransmitters, pages 217–225
©1981 Alan R. Liss, Inc., 150 Fifth Avenue, New York, NY 10011

Comparative Aspects of Raphé-Induced Modulation of Evoked and Spontaneous Cerebellar Unit Activity

Jean C. Strahlendorf, Howard K. Strahlendorf, and Charles D. Barnes

Previous anatomical studies employing HRP and uptake of labeled leucine have demonstrated the existence in cat, monkey, and rat of direct high-density projections from superior central nuclei (CS) (nuclei raphé pontis and centralis superior), inferior central nucleus (CI) (raphé magnus), and raphé dorsalis to the cerebellar cortex and deep cerebellar nuclei [Batini et al, 1977; Bobillier et al, 1976; Frankfurter et al, 1977; Hökfelt and Fuxe, 1969; Shinnar et al, 1975]. Bloom et al [1972] concluded from data derived from studies designed to decrease the influence of cerebellar intrinsic cortical interneurons that 5-hydroxytryptamine (5-HT) acted upon postsynaptic membranes of Purkinje cells, extrapolating that these cells receive inhibitory 5-HT afferents. Morphologic fluorescent studies [Chan-Palay, 1975, 1977] have shown that serotonergic neurons arising from the various raphé nuclei terminate in the molecular and granular cell layers of the cerebellar cortex. Within the molecular layer, serotonergic fibers resemble parallel fibers that do not appear to arise from granule cells. Granule layer terminations resemble claw-like or rosette configurations characteristic of mossy fiber innervation of granule cells. The third class of extrinsic indoleamine afferents consists of diffuse branching axons to all layers lying in close apposition to Purkinje cells. This latter group appears not to establish synaptic contacts, perhaps representing a neurohumoral mode of transmission. Therefore, anatomical evidence would suggest that the serotonergic fibers may make synaptic and nonsynaptic terminations on the Purkinje cells of the cerebellar cortex. The purpose of this study was to investigate the influence of 5-HT upon identifiable target cells within the cerebellum, ie, Purkinkje cells. To this end, a comparison of the effects of raphé stimulation upon spontaneous and evoked climbing and mossy fiber inputs, as well as GABA-mediated inhibition of Purkinje cells, was made.

METHODS AND MATERIALS

Cats anesthetized with α-choralose, 60 mg/kg ip, were flaxedilized and subsequently artificially respired. Blood pressure was monitored and the mean pressure was maintained above 90 mm of Hg. Body temperature was adjusted between 37 and 39°C. A craniotomy was performed between the bony tentorium and the occipital ridge over the left side of the cerebellum and over the pericruciate region of the right cerebral cortex. Raphé nuclei were identified by stereotaxic coordinates and surface landmarks (venous sinus, occipital bone). To avoid the bony tentorium, stimulating electrodes were positioned into raphé nuclei at an angle. The cerebellum was exposed at the level of the fissura prima (which separates lobules V and V1), thereby facilitating the exploration of both anterior and posterior cerebellar vermi and out to the lateral edge of pars intermedia of the anterior lobe. To reduce movement artifacts, 4% agar in a saline solution was placed on the exposed cerebellar cortex.

Raphé sites were activated bipolarly or monopolarly by means of a concentric stimulating electrode 0.5 mm in diameter with all stimuli being monophasic constant current square waves. The ipsilateral superficial nerve of the forelimb was dissected and stimulated in a pool of warm oil. Radial nerve stimuli (1.5–2 times threshold, 2 Hz, 0.05–0.1 msec, 2–3 shocks) were delivered through a pair of platinum hook electrodes. The contralateral sensorimotor cortex was activated with monopolar stimuli (1–3 shocks, 0.1 msec, 2-msec interval, 0.4–2 mA) through a silver-ball electrode placed 6–8 mm lateral to midline and anterior to the pericruciate sulcus on the sigmoid gyrus. Parallel fiber activation for eliciting "off-beam" inhibition was achieved by an array of four concentric electrodes placed in close apposition to the recording electrode.

Extracellular action potentials of Purkinje cells were recorded with either stainless-steel microelectrodes or glass micropipets filled with 2 M NaCl saturated with fast green (4–10 MΩ impedances) angled in the region of the anterior and posterior cerebellar vermi. Purkinje cells were identified by characteristic climbing fiber response consisting of a complex discharge firing pattern [Thach, 1968].

Action potentials that fell within the upper and lower window settings of an amplitude analyzer were displayed in a raster pattern on a storage oscilloscope. A comparison of changes in evoked and spontaneous activity was made by examining identical portions of the raster tracing with and without raphé conditioning. Upon termination of each experiment, the presence of a fast green spot or Prussian blue reaction product was used to verify electrode placements in the cerebellar cortex and raphé, respectively.

RESULTS

Spontaneous and Evoked Excitation

Electrical stimulation of raphé nuclei produces a period of prolonged inhibition of spontaneously discharging Purkinje cells (sometimes lasting as long as 1600 msec) occasionally preceded by a transient degree of bursting or entrainment (Fig. 1). Superficial radial nerve stimulation elicited simple spike excitation of Purkinje cells. Purkinje cells that were both spontaneously active and displayed the simple spike discharge were tested by raphé conditioning (50 Hz, 25–500 μA, three shocks) and analyzed by the method of Hoffer et al, [1978]. The period of spontaneous activity selected for examination was 200 msec in duration, 100 msec after raphé stimulation. A previous study [Strahlendorf et al, 1979a] revealed that raphé stimulation resulted in inhibition with an onset latency less than 50 msec, lasting as long as 1600 msec. For evoked activity computations, conditioning of the raphé was set approximately 60–100 msec prior to the three types of test shocks, allowing the evoked responses to fall within the raphé-induced inhibitory phase. Discharge rates for the spontaneous period were computed by multiplying the number of spikes within the examined 200 msec by 5 and dividing by the number of raster steps, yielding spikes per second. Discharge rates for evoked activity were determined by dividing the total number of spikes

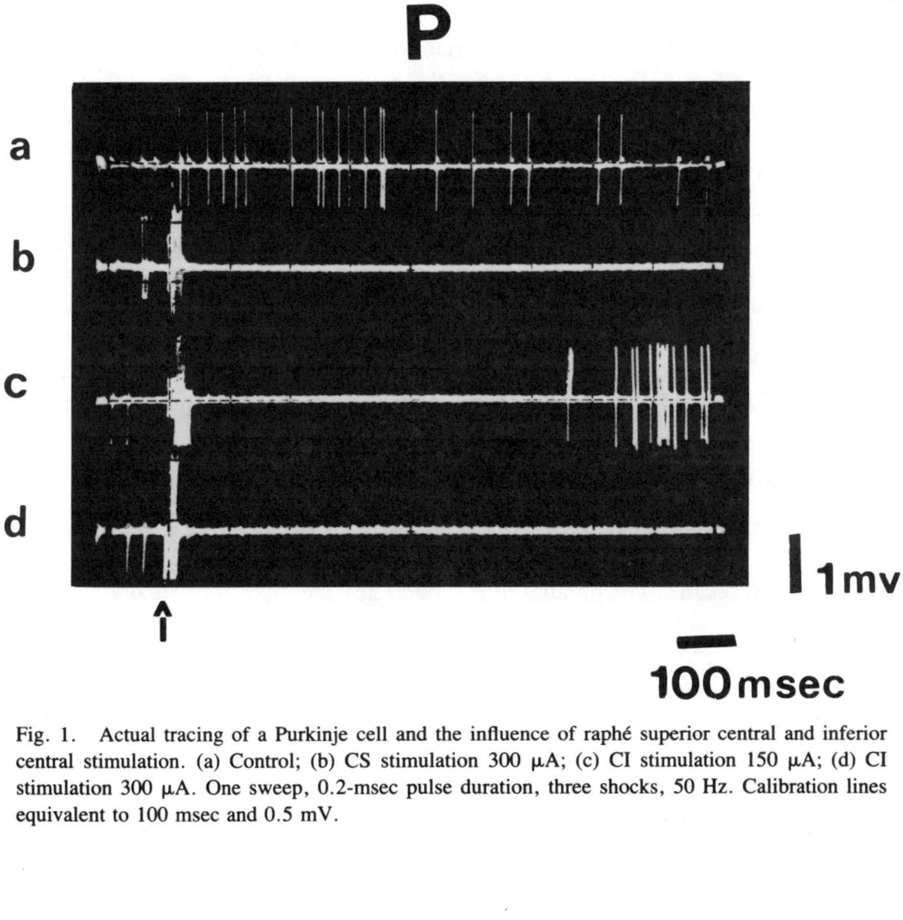

Fig. 1. Actual tracing of a Purkinje cell and the influence of raphé superior central and inferior central stimulation. (a) Control; (b) CS stimulation 300 μA; (c) CI stimulation 150 μA; (d) CI stimulation 300 μA. One sweep, 0.2-msec pulse duration, three shocks, 50 Hz. Calibration lines equivalent to 100 msec and 0.5 mV.

observed by the number of raster steps, yielding spikes per stimuli. Figure 2 depicts the preferential inhibitory influence of the raphé nuclei upon the evoked simple spike discharge as compared with spontaneous activity. Based upon the analysis of 16 Purkinje cells, 12 of these cells showed a greater inhibition of evoked, in contrast to spontaneous, activity following raphé stimulation. This calculation was based on the 12 cells that fell above the isoequal line. The average percentage of change (exclusive of the four outlying cells) of evoked and spontaneous activity was 86.1 (SE ± 3.0) and 56.1 (SE ± 4.6), respectively. The average number of spikes per stimuli before and after raphé stimulation was 3.8 and 0.6. In the four cells in which spontaneous activity was more affected than evoked activity, the average percentage of change was 69.6 and 57.1, respectively. Therefore, the influence exerted by the raphé nucleus and presumably endogenously released serotonin is most pronounced on evoked spike activity in contrast to spontaneous activity.

Figure 3 summarizes the inhibitory influence of raphé stimulation (CI and CS) on evoked activity elicited via sensorimotor cortex stimulation versus spontaneous activity. The complex spike activity fell (based on an average from 10 Purkinje cells) from 4.13 spikes per stimuli to 0.8 spikes per stimuli, representing an 86.5 (SE ± 4.6) change in comparison to a 65.1 (SE ± 9.1) decrease in spontaneous activity. In 2 of 12 cells, the number of spikes per stimuli was increased after raphé stimulation, suggestive of a potentiation of evoked complex activation. In the three cells in which spontaneous activity was more affected than evoked activity, the average percentage of change was 68.7 and 43.3, respectively. In 10 Purkinje cells, raphé conditioning of only the evoked simple and complex activity revealed a 75.4% (SE ± 7.7) inhibition. In another series of experiments, the inferior central nucleus was stimulated at current strengths that altered evoked activity but failed to influence spontaneous activity. Radial- and somatosensory cortex-derived discharges of Purkinje cells showed a 53.1% decrease following raphé stimulation. As previously observed, the influence exerted by the raphé nuclei was more apparent on evoked spike activity in comparison to spontaneous activity.

Evoked Inhibition: "Off-Beam" Inhibition

Inhibition of Purkinje cells that is not preceded by excitation and has a latency of onset 2–3 msec longer than expected for an exitation evoked by mossy fibers is termed "off beam." The parallel fibers mediating the input activate only basket cells that in turn send their axons laterally to inhibit Purkinje cells [Eccles et al, 1967]. Of 20 Purkinje cells tested, 16 showed that the typical "off-beam" inhibition was dramatically influenced by CI conditioning. The average amperage used was 30 μA (two shocks, 50 Hz, 0.1 msec), a value that caused a minimal influence on spontaneous activity (15.8 ± 4.2, n = 16), but caused a long-term augmentation (48.8% ± 4.1) of basket–stellate cell activation. Specifically, CI

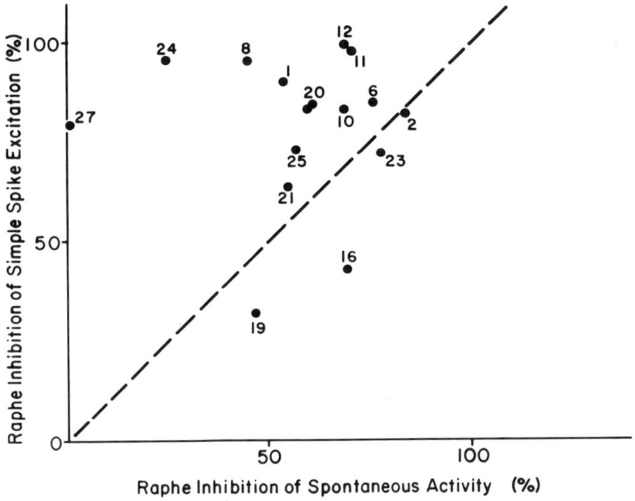

Fig. 2. The graph similar to that depicted by Hoffer et al [1978] summarizes the inhibitory influence of raphé stimulation (50–200 μA, 1–3 shocks, 0.1–0.2 msec, 50 Hz) on evoked activity elicited via superficial radial nerve stimulation versus spontaneous activity. The average percentage of change in evoked and spontaneous activity was 86.1 (SE ± 3.0) and 56.1 (SE ± 4.6), respectively. The dashed line represents theoretical values if the raphé effect were equally potent on evoked as well as spontaneous activity. The influence exerted by the raphé nucleus (CI) or endogenously released serotonin is most pronounced on evoked spike activity in contrast to spontaneous activity.

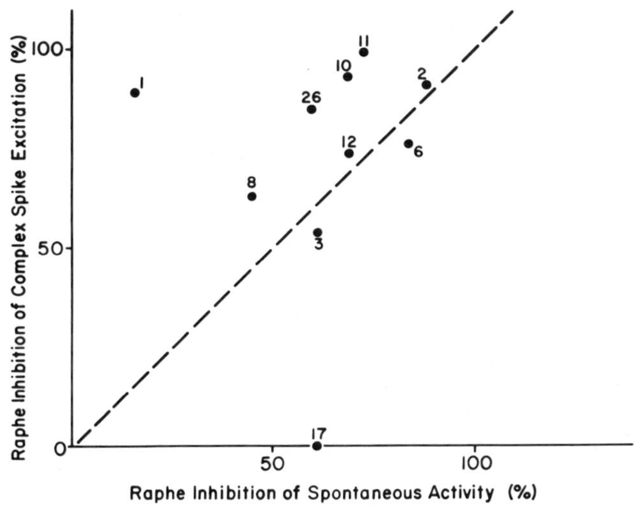

Fig. 3. The graph summarizes the inhibitory influence of raphé stimulation on evoked activity elicited via sensorimotor cortex stimulation versus spontaneous activity. There was a 86.5% change in evoked activity in comparison to a 65.1% decrease in spontaneous activity. As previously observed, the influence exerted by CI was more apparent on evoked spike activity than on spontaneous activity.

conditioning elicited a long-term inhibition following parallel fiber activation (see Methods for experimental protocol). The percentages were based upon the number of Purkinje cell discharges occurring 500 msec after parallel fiber activation with and without raphé conditioning (Fig. 4).

DISCUSSION

The findings of the present study reveal a preferential inhibitory influence of the raphé nuclei on evoked simple and complex activity rather than spontaneous firing of Purkinje cells. Furthermore, the raphé nuclei appear to augment GABAergic "off-beam" inhibition of Purkinje cells. Previous work in our laboratories [Strahlendorf et al, 1979a] has helped affirm the serotonergic nature of these responses. Raphé-induced depressant effects were diminished following the administration of the purported 5-HT antagonist methysergide. Furthermore, this agent often enhanced the spontaneous discharge of Purkinje cells suggesting that the serotonergic system exerts a tonic inhibitory influence on Purkinje cells. The serotonin precursor 1-tryptophan intensified raphé-induced inhibition and when administered subsequent to methysergide 1-tryptophan reversed the antagonist action.

Anatomical studies indicate similarities in ultrastructural morphology between the raphé nuclei and the locus coeruleus terminations in the various layers of the cerebellar cortex [Chan-Palay, 1977]. In addition to establishing synaptic contacts within the cerebellar cortex, axons of the locus coeruleus and raphé nuclei terminate nonsynaptically within this area, ie, morphologically identifiable junctions between pre- and postsynaptic elements are not defined. Norepinephrine and serotonin present in the nonsynaptic axon may be considered neurohumoral agents, with a more general dispersion into the neural mileu. Previous investigators have shown that locus coeruleus activation or iontophoretic application of norepinephrine predominantly inhibits spontaneous firing of Purkinje cells. A comparison of evoked (complex and simple activation) and spontaneous activities has revealed that the latter is preferentially inhibited by the noradrenergic system [Hoffer et al, 1978]. Based upon these findings, this nucleus appears to increase the "signal-to-noise" ratio, with the spontaneous activity representing the noise component and the evoked activity representing the signal. By contrast, the serotonergic inhibition is predominantly imposed on evoked versus spontaneous discharges. The analogous conclusion appears to be that the raphé nuclei may decrease or attenuate the "signal-to-noise" ratio, preferentially diminishing the numerator. The physiologic significance of these opposing regulators of cerebellar function is as yet unclear. Interestingly, it has recently been shown that 5-HT appears to exert a preferential inhibition of evoked versus spontaneous activity within the cerebral cortex [Woodward et al, 1980]. Analogous to the cerebellum, the norepinephrine and serotonin fibers innervate the cerebral cortex in a very similar manner [Lidov et al, 1980].

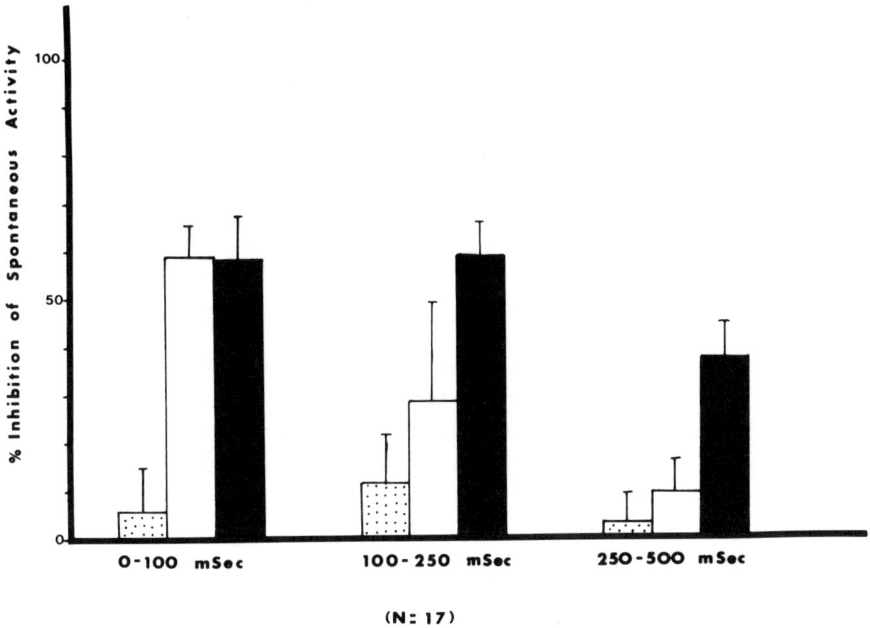

Fig. 4. These histograms display the influence of raphé stimulation (CI) on spontaneous activity and "off-beam" inhibition elicited by an array of four electrodes placed on the cerebellar surface. The X-axis represents time after test shock was applied to the surface of the cerebellum for eliciting "off-beam" inhibition. The Y-axis represents the percentage of inhibition of Purkinje cell firing. Stippled bars, effect of raphé stimulation on spontaneous activity. It should be noted that strengths of stimuli delivered to the raphé were chosen to minimally affect the spontaneous activity. Clear bars, effect of "off-beam" inhibition alone on Purkinje cells. Dark bars, effect of "off-beam" inhibition on Purkinje cells preceded 100 msec by raphé stimulation.

Since the predominant effect of raphé stimulation of Purkinje cells is that of inhibition, and since GABA is the purported mediator of many inhibitory synapses in the cerebellum (stellate, basket, and Golgi cells), previous experiments examined the possibility that raphé-induced inhibition invoked intrinsic GABAergic mechanisms. These studies showed that systemically administered diazepam, a potentiator of GABA activity, potentiated raphé inhibition, whereas bicuculline reversed the diazepam augmentation and given alone diminished raphé-elicited inhibition of Purkinje cells [Strahlendorf et al, 1979b]. The present study revealed that raphé conditioning prevented the normal return of Purkinje cell activity following "off-beam" inhibition, indicating a possible 5-HT-mediated long-term augmentation of basket–stellate cell elicited suppression. Since GABA is the presumed transmitter for these cortical interneurons, endogenously released 5-HT may be prolonging GABA-mediated inhibition. The effects of the raphé and presumably serotonin may be expressed through actions on pre- or postsynaptic mechanisms. The synaptic efficacy for GABA can be enhanced by a number of mechanisms: (1) facilitation of GABA release; (2) blockade of GABA reuptake; or (3) increased responsiveness of postsynaptic membranes to GABA. Anatomical evidence provides the framework for such an interaction within the cerebellar cortex. Specifically, Chan-Palay [1977] has revealed that indoleamine fibers terminate on axons of basket and stellate cells. Further experiments will be designed to detail further the most likely explanation underlying the interaction between 5-HT and GABA within the cerebellar cortex.

SUMMARY

Electrical activation of the raphé complex predominantly inhibits spontaneous firing of randomly encountered Purkinje cells. These experiments were performed to compare the effects of raphé stimulation upon spontaneous and evoked climbing and mossy fiber inputs, as well as GABA-mediated inhibition of Purkinje cells. The findings revealed a preferential inhibitory influence of the raphé nuclei on evoked simple and complex activity rather than spontaneous firing of Purkinje cells. Furthermore, the raphé appears to augment GABAergic "off-beam" inhibition of Purkinje cells.

ACKNOWLEDGMENTS

This work was supported by NIH Grant HL 7289 and by the Tarbox Parkinson's Disease Institute.

REFERENCES

1. Batini C, Corvisier J, Hardy O: Projections of bulbar reticular nuclei and raphé nuclei to lobules VI and VII of cat cerebellar cortex. Compt Rend Hebdoma Seances Acad Sci D Sci Nat 284:1805–1806, 1977.
2. Bloom FE, Hoffer BJ, Siggens GR, Barker JL, Nicoll RA: Effects of serotonin on central neurons: Microiontophoretic administration. Fed Pro 31:97–106, 1972.
3. Bobillier P, Seguin S, Petitjean F, Salvert D, Touret M, Jouvet M: The raphé nuclei of the cat brain stem: A topographical atlas of their efferent projections as revealed by autoradiography. Brain Res 113:449–486, 1976.
4. Chan-Palay V: Fine structure of labeled axons in the cerebellar cortex and nuclei of rodents and primates after intraventricular injections with tritiated serotonin. Anat Embryol 148:235–265, 1975.
5. Chan-Palay V: "Cerebellar Dentate Nucleus: Organization, Cytology and Transmitters." Berlin: Springer–Verlag, 1977, p 548.
6. Eccles JC, Ito M, Szentagothai J: "The Cerebellum as a Neuronal Machine." New York: Springer, 1967, p 156.
7. Frankfurter A, Weber JT, Harting JK: Brainstem projections to lobule VII of posterior vermis in the squirrel monkey: As demonstrated by the retrograde axonal transport of tritiated horseradish peroxidase. Brain Res 124:135–139, 1977.
8. Hoffer BJ, Freedman R, Woodward DJ, Puro D, Moises H: A functional role for adrenergic input to the cerebellar cortex: Interaction of norepinephrine with mossy and climbing fiber

excitation and GABA-mediated inhibition In Garattini S, Pujol TF, and Samanin R (eds): "Interactions between putative neurotransmitters in the brain." New York: Raven Press, 1978, pp 231–243.

9. Hökfelt T, Fuxe K: Cerebellar monoamine nerve terminals: A new type of afferent fibers to the cortex cerebelli. Exp Brain Res 9:63–72, 1969.
10. Lidov HGW, Grzanna R, Molliver ME: The serotonin innervation of the cerebral cortex in the rat—An immunohistochemical analysis. Neuroscience 5:207–227, 1980.
11. Shinnar S, Maciewicz RJ, Shofer RJ: A raphé projection to cat cerebellar cortex. Brain Res 97:139–143, 1975.
12. Strahlendorf JC, Strahlendorf HK, Barnes CD: Modulation of cerebellar neuronal activity by raphé stimulation. Brain Res 169:565–569, 1979a.
13. Strahlendorf JC, Strahlendorf HK, Barnes CD: A possible interplay of GABA with serotonergic and noradrenergic systems in the cerebellum. Proc Pharmacol Soc 22:219–222, 1979b.
14. Thach WT Jr: Discharge of Purkinje and cerebellar nuclear neurons during rapidly alternating arm movements in the monkey. J Neurophysiol 31:785–797, 1968.
15. Woodward DJ, Waterhouse BD, Moises HC: Serotonergic suppression of somatosensory cortical neuronal responses to putative neurotransmitters and afferent synaptic inputs. Neurosc Abst 6:448, 1980.

Index

PROGRESS IN CLINICAL AND BIOLOGICAL RESEARCH

Series Editors
Nathan Back
George J. Brewer
Vincent P. Eijsvoogel
Robert Grover
Kurt Hirschhorn
Seymour S. Kety
Sidney Udenfriend
Jonathan W. Uhr

Vol 1: **Erythrocyte Structure and Function,** George J. Brewer, *Editor*

Vol 2: **Preventability of Perinatal Injury,** Karlis Adamsons and Howard A. Fox, *Editors*

Vol 3: **Infections of the Fetus and the Newborn Infant,** Saul Krugman and Anne A. Gershon, *Editors*

Vol 4: **Conflicts in Childhood Cancer: An Evaluation of Current Management,** Lucius F. Sinks and John O. Godden, *Editors*

Vol 5: **Trace Components of Plasma: Isolation and Clinical Significance,** G.A. Jamieson and T.J. Greenwalt, *Editors*

Vol 6: **Prostatic Disease,** H. Marberger, H. Haschek, H.K.A. Schirmer, J.A.C. Colston, and E. Witkin, *Editors*

Vol 7: **Blood Pressure, Edema and Proteinuria in Pregnancy,** Emanuel A. Friedman, *Editor*

Vol 8: **Cell Surface Receptors,** Garth L. Nicolson, Michael A. Raftery, Martin Rodbell, and C. Fred Fox, *Editors*

Vol 9: **Membranes and Neoplasia: New Approaches and Strategies,** Vincent T. Marchesi, *Editor*

Vol 10: **Diabetes and Other Endocrine Disorders During Pregnancy and in the Newborn,** Maria I. New and Robert H. Fiser, *Editors*

Vol 11: **Clinical Uses of Frozen-Thawed Red Blood Cells,** John A. Griep, *Editor*

Vol 12: **Breast Cancer,** Albert C.W. Montague, Geary L. Stonesifer, Jr., and Edward F. Lewison, *Editors*

Vol 13: **The Granulocyte: Function and Clinical Utilization,** Tibor J. Greenwalt and G.A. Jamieson, *Editors*

Vol 14: **Zinc Metabolism: Current Aspects in Health and Disease,** George J. Brewer and Ananda S. Prasad, *Editors*

Vol 15: **Cellular Neurobiology,** Zach Hall, Regis Kelly, and C. Fred Fox, *Editors*

Vol 16: **HLA and Malignancy,** Gerald P. Murphy, *Editor*

Vol 17: **Cell Shape and Surface Architecture,** Jean Paul Revel, Ulf Henning, and C. Fred Fox, *Editors*

Vol 18: **Tay-Sachs Disease: Screening and Prevention,** Michael M. Kaback, *Editor*

Vol 19: **Blood Substitutes and Plasma Expanders,** G.A. Jamieson and T.J. Greenwalt, *Editors*

Vol 20: **Erythrocyte Membranes: Recent Clinical and Experimental Advances,** Walter C. Kruckeberg, John W. Eaton, and George J. Brewer, *Editors*

Vol 21: **The Red Cell,** George J. Brewer, *Editor*

Vol 22: **Molecular Aspects of Membrane Transport,** Dale Oxender and C. Fred Fox, *Editors*

Vol 23: **Cell Surface Carbohydrates and Biological Recognition,** Vincent T. Marchesi, Victor Ginsburg, Phillips W. Robbins, and C. Fred Fox, *Editors*

Vol 24: **Twin Research, Proceedings of the Second International Congress on Twin Studies,** Walter E. Nance, *Editor*

Published in 3 Volumes:
Part A: **Psychology and Methodology**
Part B: **Biology and Epidemiology**
Part C: **Clinical Studies**

Vol 25: **Recent Advances in Clinical Oncology,** Tapan A. Hazra and Michael C. Beachley, *Editors*

Vol 26: **Origin and Natural History of Cell Lines,** Claudio Barigozzi, *Editor*

Vol 27: **Membrane Mechanisms of Drugs of Abuse,** Charles W. Sharp and Leo G. Abood, *Editors*

Vol 28: **The Blood Platelet in Transfusion Therapy,** G.A. Jamieson and Tibor J. Greenwalt, *Editors*

Vol 29: **Biomedical Applications of the Horseshoe Crab (Limulidae),** Elias Cohen, *Editor-in-Chief*

Vol 30: **Normal and Abnormal Red Cell Membranes,** Samuel E. Lux, Vincent T. Marchesi, and C. Fred Fox, *Editors*

Vol 31: **Transmembrane Signaling,** Mark Bitensky, R. John Collier, Donald F. Steiner, and C. Fred Fox, *Editors*

Vol 32: **Genetic Analysis of Common Diseases: Applications to Predictive Factors in Coronary Disease,** Charles F. Sing and Mark Skolnick, *Editors*

Vol 33: **Prostate Cancer and Hormone Receptors,** Gerald P. Murphy and Avery A. Sandberg, *Editors*

Vol 34: **The Management of Genetic Disorders,** Constantine J. Papadatos and Christos S. Bartsocas, *Editors*

Vol 35: **Antibiotics and Hospitals,** Carlo Grassi and Giuseppe Ostino, *Editors*

Vol 36: **Drug and Chemical Risks to the Fetus and Newborn,** Richard H. Schwarz and Sumner J. Yaffe, *Editors*

Vol 37: **Models for Prostate Cancer,** Gerald P. Murphy, *Editor*

Vol 38: **Ethics, Humanism, and Medicine,** Marc D. Basson, *Editor*

Vol 39: **Neurochemistry and Clinical Neurology,** Leontino Battistin, George Hashim, and Abel Lajtha, *Editors*

Vol 40: **Biological Recognition and Assembly,** David S. Eisenberg, James A. Lake, and C. Fred Fox, *Editors*

Vol 41: **Tumor Cell Surfaces and Malignancy,** Richard O. Hynes and C. Fred Fox, *Editors*

Vol 42: **Membranes, Receptors, and the Immune Response: 80 Years After Ehrlich's Side Chain Theory,** Edward P. Cohen and Heinz Köhler, *Editors*

Vol 43: **Immunobiology of the Erythrocyte,** S. Gerald Sandler, Jacob Nusbacher, and Moses S. Schanfield, *Editors*

Vol 44: **Perinatal Medicine Today,** Bruce K. Young, *Editor*

Vol 45: **Mammalian Genetics and Cancer: The Jackson Laboratory Fiftieth Anniversary Symposium,** Elizabeth S. Russell, *Editor*

Vol 46: **Etiology of Cleft Lip and Cleft Palate,** Michael Melnick, David Bixler, and Edward D. Shields, *Editors*

Vol 47: **New Developments With Human and Veterinary Vaccines,** A. Mizrahi, I. Hertman, M.A. Klingberg, and A. Kohn, *Editors*

Vol 48: **Cloning of Human Tumor Stem Cells,** Sydney E. Salmon, *Editor*

Vol 49: **Myelin: Chemistry and Biology,** George A. Hashim, *Editor*